Anthropocene: A Very Short Introduction

VERY SHORT INTRODUCTIONS are for anyone wanting a stimulating and accessible way into a new subject. They are written by experts, and have been translated into more than 45 different languages.

The series began in 1995, and now covers a wide variety of topics in every discipline. The VSI library currently contains over 550 volumes—a Very Short Introduction to everything from Psychology and Philosophy of Science to American History and Relativity—and continues to grow in every subject area.

Very Short Introductions available now:

Available soon:

For more information visit our website

www.oup.com/vsi/

Erle C. Ellis

ANTHROPOCENE

A Very Short Introduction

OXFORD

UNIVERSITY PRESS

OXFORD
UNIVERSITY PRESS

Great Clarendon Street, Oxford, OX2 6DP,
United Kingdom

Oxford University Press is a department of the University of Oxford.
It furthers the University's objective of excellence in research, scholarship,
and education by publishing worldwide. Oxford is a registered trade mark of
Oxford University Press in the UK and in certain other countries

© Erle C. Ellis 2018

The moral rights of the author have been asserted

First edition published in 2018

Impression: 6

Published in the United States of America by Oxford University Press
198 Madison Avenue, New York, NY 10016, United States of America

British Library Cataloguing in Publication Data
Data available

Library of Congress Control Number: 2017959738

ISBN 978-0-19-879298-7

Printed and bound by CPI Group (UK) Ltd, Croydon, CR0 4YY

Contents

Preface

Rewriting history is an ambitious project. Even more so when this involves an entire planet and features a new leading actor. But that is precisely what this book is about.

The history of your planet and your role in it is being rewritten to include a new chapter; a chapter in which you play a leading role. We humans, the Anthropos, have so greatly altered Earth's functioning that scientists now propose to recognize this with a new interval of geologic time: the Anthropocene. Unlike prior geologic times, the proposal to mark an interval in which humans have become a 'great force of nature' has exploded across the scholarly world and beyond.

The future of the Anthropocene remains unsettled. Scientific debate still swirls among the various proposals to define an 'age of humans', including the option of rejecting the Anthropocene outright. As a work in progress, no book can provide the final word on what the Anthropocene is or will become. My goal here is simpler; to provide you with the background needed to understand the Anthropocene as a scientific proposal and to explain why it has become so broadly influential. In the process, I hope you will become as inspired as I am to more consciously and proactively shape a better future for the age of humans.

Acknowledgements

I could not have written this book without the support of my wife, Ariane de Bremond. My father, Robert Ellis, encouraged me to do the book, and my children, Ryan and Amaia, kept me inspired. I am indebted to Matthew Edgeworth, Martin Head, Peter Kareiva, Laura Martin, John McNeill, Will Steffen, Chris Thomas, Zev Trachtenberg, and Alex Wolfe for reviewing chapters and providing invaluable advice on improving them. Discussions and advice from Mark Maslin, Tim Lenton, and Andrew Bauer moved things along and helped with reviews, and Jared Margulies, Adam Dixon, and Jason Chang provided helpful feedback. My colleagues in the Anthropocene Working Group of the Subcommission of Quaternary Stratigraphy of the International Commission on Stratigraphy have shaped my thinking in so many ways. My thanks to all of them, especially Jan Zalasiewicz, Colin Waters, and Mark Williams, who made me welcome, even as my views on Anthropocene formalization within the Geologic Time Scale have differed from the consensus view. Latha Menon provided invaluable editorial advice and Jenny Nugee made sure everything was in order. This book was written on leave from my position at the University of Maryland, Baltimore County, with resources provided by the Centre for Development and Environment at the University of Bern in Switzerland. I am grateful to both for their support.

List of illustrations

Chapter 1
Origins

'We are in the Anthropocene!' exclaimed Nobel-prize winning atmospheric chemist Paul Crutzen in frustration at a conference in 2000. Why were his colleagues still calling our time the Holocene? Humans had so clearly reshaped Earth since the last ice age ended, the beginning of the Holocene Epoch. From this moment on, the proposal to rename Earth's current interval of geological time after us, the *Anthropos*, has been gaining extraordinary traction—and critics—both inside and outside the academy.

Why did such an esoteric geologic term rise so quickly to become a flashpoint of scholarly discussion and a popular phenomenon around the world? To understand this, it will help to look deeper, beyond the science, into the origin stories told across human societies since time began.

From prehistory to present, the human role in nature—as progeny, partner, steward, gardener, or destroyer—has been defined and redefined by narratives explaining human emergence on Earth. Origin stories gave humans a privileged place at the centre of divine creation in the Abrahamic religions. Copernicus and Darwin built new narratives from scientific evidence and humans became just another animal on just another planet orbiting just another ordinary star.

The Anthropocene demands an even greater adjustment of our perspectives. As geologists and others struggle for and against various proposals to formalize the Anthropocene, it should come as no surprise that their efforts have become entangled with both age-old worldviews and contemporary debates on the role of humans in nature and even what it means to be human.

A great force of nature

Crutzen's outburst was rooted in his experience investigating human-caused changes in Earth's atmosphere and their profound global consequences: the hole in Earth's protective ozone layer and global climate change. To hear his colleagues speak of Earth's current state without reference to these profound anthropogenic changes was just too much to bear. It was time to accept that the relatively stable conditions of the Holocene Epoch were over.

Crutzen was not alone. Ecologist Eugene Stoermer had been using the term Anthropocene informally with students and colleagues since the 1980s. In 2000, the two published a brief note in a scientific newsletter, the first formal appearance of the term in print—though *New York Times* writer Andy Revkin had used 'Anthrocene' in his book on climate change in 1992. In this first publication, Crutzen and Stoermer linked the Anthropocene with carbon dioxide emissions from fossil fuel combustion and dated it to the start of the Industrial Revolution at the end of the 18th century. In so doing, they were building on a mass of earlier work describing anthropogenic environmental changes. With Crutzen's proposal, these many threads had finally come together in a proposal to mark human emergence as a 'great force of nature' in the historical record of Earth.

Changing history

Overwhelming evidence now confirms that humans are changing Earth in unprecedented ways. Global climate change, acidifying oceans, shifting global cycles of carbon, nitrogen, and other elements, forests and other natural habitats transformed into farms and cities, widespread pollution, radioactive fallout, plastic accumulation, the course of rivers altered, mass extinction of species, human transport and introduction of species around the world. These are just some of the many different human-induced global environmental changes that will most likely leave a lasting record in rock: the basis for marking new intervals of geologic time.

With such an abundance of evidence, the proposal to recognize the Anthropocene as a new interval of geologic time, the Anthropocene epoch, would seem without issue. Yet the opposite is true. The Anthropocene remains highly controversial even among Earth scientists. Arguments rage over whether there are sufficient scientific grounds to recognize such a comparatively brief and novel epoch, while others argue over the optimal timing and evidence. Proposals for the start of the Anthropocene range from early human control of fire, to the rise of agriculture more than 10,000 years ago, to the peak year of nuclear fallout in 1964, supported by evidence ranging from gas bubbles trapped in ice cores and widespread deposits of soot and radionuclides, to the appearance of domesticated maize pollen in sediment cores around the world. And this is just to scrape the surface of the many disputes instigated by the Anthropocene proposal.

The proposal to rename our time the 'age of humans' has probably been even more disruptive outside the Earth sciences, kindling intense debates, sustained discussions, and transformative new research in disciplines as wide ranging as philosophy and archaeology, anthropology, geography, history, engineering, ecology, design, law, the arts, and political science. The Anthropocene debate has even boiled over into the media and across the public realm, from the water cooler to popular music. Does an age of humans mean the end of nature? Who is responsible for the Anthropocene? *Homo sapiens*? The first farmers? Wealthy consumers of the industrial age? And is the Anthropocene necessarily a catastrophe—an environmental disaster and the end of humanity—or could there be a 'good Anthropocene' in which both humans and nature might thrive together into the deep future?

The many heated controversies surrounding the Anthropocene make clear that there is far more at stake than just a new interval of geologic time. The significance of the Anthropocene resides in its role as a new lens through which age-old narratives and philosophical questions are being revisited and rewritten. The Anthropocene is both a new narrative relating humans and nature and a bold new scientific paradigm—a 'Second Copernican Revolution'—with the potential to radically revise the way we think of what it means to be human.

Origin stories

Human societies have always used narratives to explain the origins of their people and their relationships to the world and its many actors—from animals and plants to more mystical beings and forces. For the ancient Greeks, Earth, as goddess Gaia, emerged from the void and gave birth to all of life and to the progenitors of their many gods, from Athena to Zeus. The Greeks' mortal human ancestors appear only after several earlier human races are created, found wanting, and sent away by the gods. Or, in a separate origin story, the god Prometheus creates humans out of clay and enables them to thrive by giving them fire stolen from the gods. The message is clear. Earth, as Gaia, creates and sustains all of nature, including the ever-battling forces of the gods. Humans are a sideshow in the mythology of ancient Greece, lucky to exist at all, and thrive only with the help of Prometheus' gift of fire. As we shall see, both Gaia and Prometheus play key roles in the origin stories of the Anthropocene.

In the first story of the Hebrew Genesis, a single all-powerful god creates cosmos, Earth, and humans in an orderly sequence. In the second, man is created first, then nature—the garden of Eden—and then woman. Their lives are effortless until tempted by a 'tree of knowledge'. An angry God then expels them from Eden, thereby forcing them and their descendants to forever cultivate the Earth to survive. Through this narrative, we learn why humans, despite their privileged central role in God's creation, are nevertheless committed, following the Fall, to toil in cultivating the Earth.

Through storylines connecting cosmos, Earth, and people with all the other actors and forces they must interact with, origin stories tell us who we are, where we came from, the role we play on Earth, and our relationships to the rest of nature. Similarly, the Anthropocene presents an account of a planet reshaped by humans. But how and why did humans become planet shapers? The Anthropocene demands an explanation.

23 October 4004 BC

At six p.m. on 23 October 2004, scientists at the Geological Society of London raised a toast to Archbishop James Ussher of Armagh. According to Bishop Ussher, 23 October 4004 BC was the precise date and moment of creation. Based on his dating, made in 1650, the universe was exactly 6,008 years old. While these connoisseurs of geologic time were certainly just having fun, it is telling that they would celebrate such a dramatically obsolete chronology of the universe. Ussher's precision might provoke laughter today, but its purpose is no less clear: to confer greater certainty on his origin story.

Even before the rise of Western scientific methods, precise chronologies of key events in Earth and human history were produced through careful analysis of trusted evidence. Bishop Ussher used the Bible to produce his chronological narrative.

Generational history (e.g. Jacob begat Joseph) and dated events (e.g. destruction of the Temple in Jerusalem) were laboriously compiled and creatively computed to produce a precise chronology connecting cosmos, Earth, human origins, and the history of Western society. Many other societies, including the Maya and Hindu, also produced detailed chronologies connecting the formation of the cosmos with human history, relying in part on painstaking astronomical observations. This widespread investment of specialized expertise in producing detailed chronologies confirms their social utility long before the rise of Western science, as a way of conferring authority on the institutions and experts that created and maintained them.

Contemporary scientific efforts have developed the most elaborate, precise, systematic, and verifiable origin story ever, linking cosmos, Earth, life, and human history within a single, increasingly detailed, and continuously improving chronology. Yet even now, many traditional, religious, and even secular communities continue to maintain their own separate and competing origin stories that contrast sharply with scientific evidence, often in the face of substantial societal pressures. For example, some still perpetuate the 'young Earth' chronology of Bishop Ussher.

The main reason for this rejection should be clear. By redefining the roles and relations of humans, Earth, and cosmos, the origin story of contemporary science challenges some of the most deeply held traditional beliefs of societies around the world. There is no role for an all-powerful God or any other mystical force. Humans play no central role in the universe. The Anthropocene goes even further, not only by confronting these traditional beliefs, but also by revising the classic origin story of contemporary science. In the Anthropocene, humans are put back into a central role on Earth, as planet shapers.

The first Copernican revolution

On 4 June 1539, Martin Luther discussed with his disciples 'a certain new astrologer who wanted to prove that the Earth moves and not the sky, the Sun, and the Moon'. The astrologer was Nicolaus Copernicus and his heliocentric theory would ultimately displace Earth from the centre of the universe.

For millennia, the only acceptable origin story of the Western world centred on Earth and began with its creation by a Christian God. The literal truth of the biblical origin story depended on this geocentric view. Unsurprisingly, efforts to displace Earth and humanity from the centre of the cosmos were resisted. It took more than a century, and the work of Tycho Brahe, Johannes Kepler, Galileo Galilei, and ultimately Isaac Newton for the Copernican revolution to succeed. But it did. By the late 17th century, at least among the Western scientific intelligentsia, Earth was no longer the centre of the universe and the need for a new origin story for Earth and the cosmos became clearer.

Layers of time

As long as a century after Ussher released his chronology, scholars such as Isaac Newton still believed that Earth was no more than 6,000 years old. The first challenge came from French naturalist Georges-Louis Leclerc, Comte de Buffon (1707–88), who published estimates that Earth was 74,000 years old in the late 18th century. His estimate was soon ridiculed and retracted under pressure, even though he actually believed Earth was even older, perhaps millions of years old.

The scientific basis for dating intervals of geologic time emerged with the discovery that distinctive banded patterns of materials and fossil creatures observed in exposed rocks and sediments could be organized into a system of horizontal layers—'strata'—formed one on top of the other. By the early 19th century, geologists had established the science of stratigraphy. Charles Lyell published his *Elements of Geology* in 1838, organizing the major stratigraphic layers identified by others into sequential intervals of time, and linked these with principles of continuous gradual change that could enable the duration of these intervals to be estimated. Putting these together in 1867, he made one of the first scientifically based estimates of Earth's age, 240 million years. His contemporaries, including Lord Kelvin, computed similar estimates that began to demolish the notion of a much younger Earth, paving the way for an entirely new origin story for the cosmos, Earth, and people.

Naked Ape

Just as geologists were revising Earth's position in the cosmic order of time, biologists were rethinking the origins of life and humans. And their central problem was time—the need for lots of it.

Charles Darwin was a close follower of geology, especially Lyell's work, and Lyell mentored Darwin following his voyage on the *Beagle*. He was invited to share his work at the Geological Society of London and was soon elected to its governing Council. But Darwin's overwhelming interest was to understand why 'one species does change into another'. In 1837, Darwin sketched this process as the branching of a single family tree. Yet it would take Darwin nearly twenty years and the fear of being scooped by Alfred Russel Wallace before he finally published his theory of evolution by natural selection in 1859.

It might seem strange to wait so long to publish one of the most important discoveries of all time. But Darwin had good reasons. As a religious man, Darwin was well aware of the controversy his theory would ignite. The claim that species originated over time through evolution—no divine act required—would not reconcile easily with the origin story of Genesis. He worked for years to strengthen it.

To confirm his theory of evolution by natural selection, Darwin needed three things. He needed evidence that species did not last for ever and that new species came into being after others. The fossil record of geology confirmed that. He needed pressures and processes that would shape species into new forms. Malthus' theory of resource limits to population growth provided the pressure—not all individuals could survive the competition for limiting resources. His study of animal and plant breeding—artificial selection—demonstrated that selective pressures could produce vastly different races, breeds, and varieties from populations of a single species. But most importantly, Darwin needed time.

Without huge spans of geologic time, hundreds of millions of years, there was no way to explain how Earth's myriad species could have arisen through natural selection alone. Fortunately, geologists would soon estimate Earth's age at hundreds of millions and ultimately billions of years. Reception for Darwin's theory continued to gain strength. In 1871, he took this a step further, focusing evolutionary theory on the story of human origins in the *Descent of Man*. Human origins were no different from those of any other animal. Our story was that of a 'Naked Ape' descended from other apes over a very long span of geologic time. Through Darwin's theory of evolution by natural selection, a new origin story was born that connected all of life, including humans, through descent from a common ancestor within a universal 'tree of life'.

In scientific circles, geologic time soon replaced biblical time, and evolution by natural selection overturned the origin story of Genesis. A new, secular, origin story connected Earth, life, and people. As noted by Thomas H. Huxley, President of the Geological Society of London in 1869, 'Biology takes her time from geology.' And unlike the story of Genesis, humans played no special role—just one species among many others evolving in no particular direction on a changing planet.

A minor role

After Darwin, and following rapid developments in astronomy, the position of humans in the history of the universe was rewritten. The cosmos originated 13.8 billion years ago in a giant explosion—the 'Big Bang'. Earth coalesced from dust and gas billions of years later, solidifying into a planet 4.5 billion years ago. Earth, one of eight planets in an irregular orbit around a typical yellow dwarf star, was located in the spiral arm of a galaxy of more than 100 billion stars, in one of more than 100 billion galaxies, together holding about 1 billion trillion stars in a continuously expanding universe.

The first life most likely arose more than 3.8 billion years ago as bacteria, evolving into more complex single-celled organisms with a nucleus, the eukaryotes, nearly 2 billion years ago. The first multicellular organisms evolved more than a billion years ago, and the first simple animals by about 800 million years ago. Life colonized land about 480 million years ago, evolving into myriad forms—most, like the non-avian dinosaurs, lost for ever in one of five mass extinction events occurring hundreds to tens of millions of years ago. Mammals first evolved 200 million years ago, then the first primates (65 million years ago), and then the first species in our direct lineage, the genus *Homo*, about 2.8 million years ago. These early human species, the hominins, were the first to shape stone tools, control fire, and migrate out of Africa and across Eurasia. Not us.

Homo sapiens emerged among other tool-making, fire-controlling, hominin species only about 300,000 years ago. For as long as 200,000 years after that, humans showed few distinguishing features besides a less robust anatomy and a slightly smaller and differently shaped skull. There would come a time when *Homo*

Time of year	ka	Event
January 1	13,800,000	Big Bang
May 1	8,500,000	Milky Way Galaxy
September 2	4,600,000	Solar system
September 6	4,400,000	Oldest rocks
September 21	3,800,000	Life (single cells, without nuclei)
October 9	3,400,000	Photosynthesis (cyanobacteria)
October 29	2,400,000	Oxygenation of atmosphere
November 15	2,000,000	Eukaryotes (first cells with nuclei)
December 5	800,000	First multicellular organisms
December 20	450,000	Land plants
December 23	300,000	Reptiles
December 25	230,000	Dinosaurs
December 26	200,000	Mammals
December 27	150,000	Birds
December 28	130,000	Flowers
December 30	65,000	Cretaceous–Paleogene boundary. Extinction of Non-avian Dinosaurs, first Primates
December 31 14:24	12,300	Hominids
22:24	2,500	Genus *Homo*, stone tools
23:44	400	Control of fire
23:48	300	*Homo sapiens* (anatomically modern humans)
23:55	100–60	'Modern' human behaviours, including symbolic markings, long-distance trade, more complex tools and settlements
23:59:32	12	Agriculture, Holocene epoch
23:59:48	5	Bronze, First Dynasty of Egypt
23:59:49	4.5	Alphabet, Wheel
23:59:53	3	Iron
23:59:32	2	Roman Empire, Christian history, invention of 0
23:59:59	0.5	Old World, New World collision

1. The cosmic calendar. Popularized by Carl Sagan, the cosmic calendar represents the history of the cosmos, Earth, life and humans as the passage of a single year, for example, Hominids appear at 2:24 PM on the last day of the year (ka = thousands of years before present).

sapiens would shape a different way of life, spread across an entire planet—and even leave it. But only in the final seconds of the cosmic calendar (Figure 1). For most of human time on Earth, our species was just one of several in the genus *Homo*, among millions more species living on an ordinary planet orbiting a typical star in a typical galaxy in a vast universe.

Changing Earth

For most natural scientists, humans have long been a sideshow; the main stage occupied by the natural world and its processes, from physics to chemistry to biology. Compared with these 'great forces of nature' and their billions of years of unbroken history, we humans are just another animal—and a newcomer at that. But even among the scientific thinkers of Darwin's time, another view was emerging. Humans were not just another primate, but a profoundly disruptive force like no other on Earth.

Among the most prominent proponents of this view was George Perkins Marsh, whose book *Man and Nature* (1864; revised as *The Earth as Modified by Human Action* in 1874) told a different story about the relationship of humans with the natural world. Ancient human societies of the Mediterranean cleared forests and tilled the land for agriculture, dramatically changing vegetation, soils, and even climate across large regions, bringing 'the face of the earth to a desolation almost as complete as that of the moon'. Humans were a destructive force capable of changing Earth permanently for the worse. In 1873, little-known geologist Antonio Stoppani went even further by defining a new time interval based on these changes, the 'Anthropozoic era'.

As the Industrial age unfolded, ever greater demands were made for Earth's resources. Powered by the combustion of fossil fuels and connected together by global networks of trade, the scale, intensity, and extent of human activities increased dramatically. Forest clearing, the tillage of soils, mining, the construction of cities, and industrial production increasingly led to pollution of water, air, and land, and the widespread conversion of natural places into bustling human landscapes left less and less room for non-human inhabitants. But the most unprecedented demonstration that humans had become a force capable of altering Earth would come out of thin air.

The end of nature

In 1895 Svante Arrhenius, building on work by John Tyndall, demonstrated that carbon dioxide and water vapour in Earth's atmosphere trapped heat energy, in a 'greenhouse effect' that warmed Earth's surface enough to support liquid water—a prerequisite for life as we know it. Moreover, he suggested that changes over time in carbon dioxide and other 'greenhouse gases' in the atmosphere could help to explain ice ages and other long-term changes in Earth's temperature. Combustion of coal might further increase this 'greenhouse warming' of the planet. This might be a good thing, he thought, at least in cold places like his native Sweden.

More than half a century after Arrhenius, evidence confirmed that carbon dioxide from fossil fuels was indeed filling Earth's atmosphere and causing temperatures to rise. In 1965, scientists warned of the dangers of anthropogenic global warming in a report to US president Lyndon Johnson. Evidence mounted and the forecast became clearer. If current trends continued, Earth would heat up dramatically, causing massive consequences for both human societies and the natural world within a few decades. Sea levels would rise, threatening cities. Climate changes would disrupt agricultural production and displace natural habitats around the world. The scientific message was clear: human activities were driving Earth in a new and potentially catastrophic direction. Scientists called for action.

In 1988, a new scientific institution was formed to assess the risks of anthropogenic global warming, the Intergovernmental Panel on Climate Change (IPCC). They were not alone. A broad community of activists, institutions, and intellectuals had been at work for more than a century addressing the many different forms of environmental harm caused by humans. Evidence that humans were altering the most basic conditions of life on Earth became their battle cry. Anthropogenic global warming was changing everything. For some, it was time to rewrite the story of humans and nature.

In 1989, journalist Bill McKibben published *The End of Nature*, the first popular book about climate change. For McKibben, human destruction of natural environments had reached its pinnacle. Modern societies had already altered, domesticated, and controlled the world more than any before, polluting and degrading water, soil, air—and the nature of life itself. By altering the climate system, humans had taken the final step. Nature untouched by humans had now disappeared through the global reach of a human-altered climate.

A new chapter

To interpret anthropogenic climate change as the 'end of nature' might be going too far. How could a product of nature—a Naked Ape—gain the capacity to end nature itself? And if nature has indeed ended, what have we now? Yet the scientific evidence is clear. Humans are indeed changing Earth in unprecedented ways. There are good reasons to accept that a new chapter of Earth history might indeed be unfolding, with humans playing a leading role.

This is why the Anthropocene has gained such great attention. Putting humans in the role of Earth-changers revises scientific narratives about humans and nature that have been developing since Copernicus' time. And controversy has always attached to efforts to rewrite the history of Earth, life, and humanity. Even while geologists are continuing to serve as the timekeepers of Earth history, a role they have played for more than two centuries, the scientific narrative of the Anthropocene is breaking new ground. New types of questions and new forms of evidence will be needed to place humans in the role of originators of a new interval in Earth history.

No other species is recognized with its own interval of geologic time. Why did humans, alone among species, gain the capacity to transform an entire planet? When did this capacity emerge—and by what mechanism? Are all humans equally a part of this transformation? What evidence is needed to answer such questions? More broadly, what does it mean to be human when this means to be part of a global force that changes everything—even the future of an entire planet? What does nature even mean in an age of humans?

Answering these questions will require a basic understanding of Earth system processes and the human-induced changes in these that inspired the Anthropocene proposal. It will also be necessary to understand the tools, procedures, and frameworks of geologic time if we are to date these changes and place them in the formal calendar of Earth history. On this basis, we will examine the full range of proposals for dating the start of the Anthropocene, from the nuclear tests of the 1950s all the way back to the beginnings of agriculture, the origins of humans as a species, and before. We will then go further to explore the many ways that the Anthropocene proposal is reshaping the sciences, stimulating the humanities, and foregrounding the politics of life on a planet transformed by humans.

Chapter 2
Earth system

'Are humans now overwhelming the great forces of nature?'
asked Will Steffen, Paul Crutzen, and historian John McNeill
in their classic 2007 article on the Anthropocene. It was a
rhetorical question. To these authors, the answer could only
be a resounding yes.

For some it might seem grandiose, even heretical, to make such a
claim. But for Steffen, Crutzen, and other Earth system scientists,
this had been the subject of decades of research. To them, the
'great forces of nature' were no godlike powers, but rather, the
processes underpinning Earth's functioning as a complex
dynamic system.

There are good reasons why the Anthropocene proposal
originated among scientists focused on understanding Earth in
this way. To confirm that humans have altered Earth's functioning
as a system, the causal mechanisms behind these alterations must
be demonstrated. Without a robust understanding of Earth as a
system—its fundamental components, their interactions, and,
most importantly, the processes that keep the Earth system stable
or induce change—it is not possible to establish the causes of
changes in the Earth system.

Spheres and cycles

The first steps on the road to Earth system science were taken by a geologist, Eduard Suess, when he introduced the terms lithosphere, hydrosphere, and biosphere in his popular textbook of 1875, *The Face of the Earth* (*das Antlitz Der Erde*). Building on Suess's terms, Vladimir Vernadsky developed the first modern scientific model of Earth as a complex system based on dynamic interactions among the 'spheres', in his 1926 book, *The Biosphere* (Figure 2).

Vernadsky characterized Earth's functioning based on exchanges of energy and matter among the spheres, with this whole system of interactions powered by energy from the sun. The biosphere played a central role in these interactions, serving as a thin green

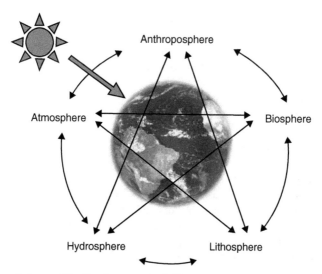

2. **Spheres of the Earth system, including an 'anthroposphere' representing the interactive global effects of human activities.**

'envelope' that regulated and enhanced these exchanges of energy and matter among Earth's atmosphere, hydrosphere, and lithosphere. By harvesting energy from the sun and carbon dioxide from the air, photosynthetic organisms gained the capacity to alter the global cycles of carbon and other elements among the spheres. And by regulating the concentrations of carbon dioxide and other greenhouse gases in the atmosphere, the biosphere permanently transformed the dynamics of Earth's climate system. Though Vernadsky is now considered the first scientist to propose that Earth's functioning as a system was transformed by the emergence of the biosphere, his work was not widely disseminated outside the Soviet Union at the time.

Gaia reborn

In the mid-1960s, Carl Sagan and other astrophysicists had a problem. Earth's climate was known to have been remarkably stable over the past 4 billion years. Yet the sun's energy output had increased by 30 per cent over this period. From the beginning, Earth was already warm enough, thanks to high concentrations of carbon dioxide in its atmosphere, to sustain liquid water and other conditions needed to support life. Why didn't Earth heat up dramatically as the sun's energy increased? Without some regulating mechanism, Earth would have become far too hot to support life as the sun heated up.

By the early 1970s, James Lovelock and Lynn Margulis had an answer. Living organisms, acting collectively as the biosphere, were themselves responsible for regulating Earth's climate and sustaining the conditions necessary to support life. Life gave birth to life itself. Gaia was reborn in the landmark hypothesis that sparked the rise of Earth system science.

The Gaia hypothesis holds that the biosphere regulates Earth's climate by acting like a thermostat. When Earth heats up, the biosphere responds by producing cooling effects. For example, organisms increase their uptake of greenhouse gases from the atmosphere and release fine particles—aerosols—into the air, helping to form clouds that reflect the sun's energy. In response to a cooling Earth, the biosphere produced opposite effects, counteracting cooling by producing warming effects—increasing greenhouse gases and reducing aerosols in the atmosphere. In this way, the biosphere could stabilize Earth's temperature through a 'negative feedback' system, countering the warming effects of increased solar energy—a process external to Earth. Negative feedbacks might also balance out the heating and cooling effects caused by processes internal to the Earth system, such as the release of greenhouse gases and aerosols by volcanoes.

The Earth system is also full of positive feedback systems, such as those controlled by Earth's global ice pack—the 'cryosphere'—when the sun melts Arctic ice. Seawater exposed to the sun is an excellent absorber of the sun's energy. Ice floating on the sea mostly reflects this energy back out into space. When the sun melts Arctic ice, it exposes more sun-absorbing seawater, enabling more heat to be absorbed. As a result, more ice will melt, causing yet more warming. In this way, sea ice melting represents a positive feedback loop, in which warming leads to yet more warming, accelerating the sun's melting of Arctic ice. At some point, this positive feedback system may reach a point of no return—a 'tipping point' or 'regime shift'—after which the melting of Arctic ice will continue until all the ice has melted.

Considering the dramatic long-term increase in the sun's energy, an Earth system unresponsive to the warming effects of the sun could not have sustained life. Some form of regulation was necessary. To understand Earth's remarkable long-term stability and ability to sustain life, this must be understood as the product of a complex system of interacting positive and negative feedbacks which shape the flows of matter and energy among the spheres.

Lovelock's Gaia hypothesis yielded a popular book and an entirely new way of thinking about life on Earth. While its proposed biospheric mechanism of long-term climate regulation has now been largely displaced by a geochemical mechanism, its most important long-term contribution has probably been its underlying framework describing Earth's functioning as a complex, dynamic system stabilized by feedback interactions among its spheres. With Gaia, climate stability in the face of a warming sun and other self-regulating behaviours came to be understood as complex, system-level processes emerging from interactions among Earth's component systems; as a whole greater than the sum of its parts. By bringing a systems approach to understanding Earth's long-term dynamics, including the concept of a dynamic biosphere and the need for computational models to incorporate dynamic interactions among Earth's spheres, Lovelock and Margulis's Gaia laid the foundations for Earth system science.

The Great Oxygenation

The biosphere's role in Earth's functioning exemplifies what it means to be a great force of nature. Scientists have long viewed Earth as a dynamic planet, at least since the work of James Hutton in the late 18th century. But Earth's transformation by the biosphere gives dynamic planet a whole new meaning. Not only did life give birth to life itself, living organisms also produced Earth's oxygen atmosphere.

Life appears to have begun as single cells in the sea about one billion years after Earth solidified as a planet. Like Venus and Mars, Earth's atmosphere then was mostly composed of carbon dioxide (CO_2). Without an oxygen (O_2) atmosphere, there was also no layer of ozone gas (O_3) derived from this O_2 to absorb the sun's life-destroying ultraviolet radiation, which inhibited the emergence of life on land.

It would be another billion years before this would start to change, thanks to the rise of organisms capable of feeding on the sun. Cells with this new biological capacity, photosynthesis, used the sun's energy to convert CO_2 and water into sugars and ultimately all of the other carbon-rich organic compounds needed to support life. As a result, a vast new supply of energy became available to support and expand the growth and development of the biosphere. And Earth's atmosphere was forever changed by the accumulation of its primary waste product: O_2 gas.

Photosynthesis changed everything. Over hundreds of millions of years, photosynthetic organisms, mostly bacteria, would fill Earth's atmosphere with O_2. This 'Great Oxygenation' of Earth's atmosphere was at first a very slow affair, as the first available O_2 reacted with iron and other minerals in Earth's oceans and crust, producing vast deposits of oxidized iron and other compounds. But once these minerals were oxidized, O_2 began to accumulate rapidly in the atmosphere. The level of atmospheric CO_2 dropped dramatically, trapped first in the carbon-rich bodies of living organisms and sequestered in deep ocean sediments as their dead bodies sank, accumulated, and ultimately formed into rocks. Organisms unsuited to a life exposed to highly reactive O_2 gas either went extinct or retreated into Earth's less oxygenated recesses.

Free oxygen in the atmosphere radically altered Earth's chemistry. Oxygen levels similar to those of today were reached during a second wave of oxygenation associated with the rise of land plants about 400 million years ago. Earth's new oxygen chemistry dissolved rocks, created new minerals, and enabled the rapid release of energy stored in organic compounds, allowing fires to burn and unlocking new forms of high-energy metabolism, such as aerobic respiration, greatly enhancing the capacity of complex, multicellular organisms to sustain themselves. In so doing, the biosphere helped produce the conditions needed for complex multicellular organisms to thrive and facilitated the appearance of life on land, shielded by a protective layer of ozone in the stratosphere. And by helping to remove and sequester CO_2, Earth's atmosphere and climate dynamics were transformed permanently, vastly reducing the greenhouse warming effects that still keep Venus' surface hot enough to melt lead. Thanks to the emergence of life, Earth's chemistry and physics were permanently altered.

Carbon and climate

The biosphere remains an active player in the Earth system today, responding to long-term changes in incoming solar energy by altering concentrations of carbon dioxide in the atmosphere. Throughout Earth history, incoming solar energy has risen and fallen as a result of cyclic changes in Earth's distance and orientation to the sun, triggering extensive glaciations at various times. In the past 2.6 million years, Earth has experienced numerous cycles of cold 'glacial' intervals or 'ice ages', during which glaciers and sea ice have crept down from the poles, interspersed with relatively warm 'interglacial' intervals, in which this ice recedes.

By drilling deep into the ice of Antarctica and Greenland, climate scientists have reconstructed these long-term cycles of Earth's temperature and atmospheric carbon dioxide through measurements made on layers of ice deposited over hundreds of thousands of years (Figure 3). Over the many glacial/interglacial cycles of this interval, atmospheric carbon dioxide rises and falls in sync with a warming and cooling Earth, partly as a result of the biosphere's active response, partly in response to carbon storage in Earth's oceans. By releasing carbon as Earth warms and taking it back up when Earth cools, the biosphere has reacted to long-term cyclic changes in incoming solar energy and warmth by amplifying these effects, forming a positive feedback system that has enhanced the dynamics of Earth's climate for more than a million years.

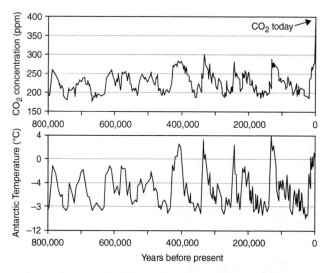

3. Changes in carbon dioxide and climate over the past 800,000 years based on ice core records from Antarctica, illustrating their correlation across glacial/interglacial cycles.

Even while the sun's energy and Earth's climate have remained relatively stable during the warm interglacial interval of the past 11,000 years, the biosphere continues to regulate the seasonal dynamics of carbon dioxide in the atmosphere within each year. Every spring, in response to the sun's warming of the northern hemisphere—where most of Earth's land and terrestrial photosynthesis are located today—the biosphere starts to take up more carbon, reducing carbon dioxide in the atmosphere. As the northern hemisphere cools down each autumn, photosynthesis slows while carbon is released from decaying vegetation, soils, and animals. These annual cycles of carbon uptake and release are part of the global cycling of carbon among the biosphere, atmosphere, and the other spheres of the Earth system forming the global 'biogeochemical' cycle of carbon, the second greatest global cycle of any element across the spheres (the oxygen cycle is more massive; Figure 4). By assessing the long-term dynamics

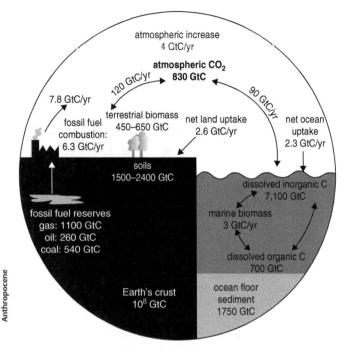

4. The global carbon cycle, in gigatonnes of carbon (GtC).

of the carbon cycle using the tools of Earth system science, observations on atmospheric carbon and temperature reveal systematic feedbacks among the biosphere, lithosphere, and atmosphere that produce both stability and instability in the face of changing solar energy inputs from the sun and other dynamic drivers of change in the Earth system.

Keeling's Curve

In March 1958, funded by the International Geophysical Year programme, Charles David Keeling hauled an infrared gas analyser to the top of Hawaii's dormant Mauna Loa volcano. His goal was to measure carbon dioxide at a remote, undisturbed site where concentrations would resemble the well-mixed conditions of Earth's atmosphere as a whole. Keeling, a young postdoctoral scientist, would soon write of his first discovery, 'witnessing for the first time nature's withdrawing CO_2 from the air for plant growth during summer and returning it each succeeding winter'. Keeling had observed the biosphere 'breathing'.

Even with this early breakthrough, Keeling's greatest contribution would require a few more years of careful measurements. In what is now known as the 'Keeling Curve', his longer-term measurements revealed a striking trend beyond the seasonal cycles of the terrestrial biosphere (Figure 5). Across years, carbon dioxide concentrations showed a distinct upward trend. In 1960, Keeling published his work, marking the first solid evidence that mass combustion of fossil fuels was in fact causing carbon dioxide to accumulate in Earth's atmosphere. By the 1970s, this continuing upward trend was gaining increasing attention from Earth scientists.

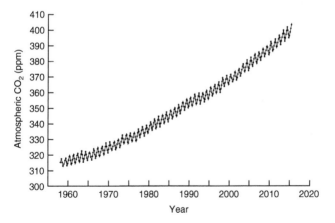

5. **The Keeling Curve. Charles David Keeling's repeated measurements of atmospheric CO_2 at Mauna Loa, Hawaii first demonstrated that CO_2 concentrations were increasing globally over time.**

Keeling's Curve demonstrated that human use of fossil fuels was rapidly changing Earth's atmosphere and potentially even the functioning of the Earth system as a whole. And atmospheric observations were only the beginning. Efforts to understand the causes and consequences of atmospheric trends in carbon dioxide would require a full accounting of the flows and cycles of carbon into and out of Earth's many reservoirs of carbon, including not only the vegetation and soils of the terrestrial biosphere and human-induced changes in these, but also Earth's oceans and even volcanic emissions—all in addition to societal emissions from the combustion of fossil fuels and the production of steel and cement. It would take an unprecedented international scientific collaboration to examine together these components of the global carbon cycle. Efforts to confirm that human societies were changing Earth's atmosphere and ultimately Earth's climate would soon stimulate the rise of a larger community of Earth system scientists.

The ozone hole

In a paper published in 1970, atmospheric chemist Paul Crutzen suggested that Earth's protective ozone layer might be threatened by emissions of a stable non-toxic gas produced naturally by bacteria living in soils. Rising to the upper levels of Earth's atmosphere, nitrous oxide gas (N_2O; also known as 'laughing gas') would be stripped apart in the harsh ultraviolet radiation of the stratosphere, where it would react with and damage the ozone layer. A reduction in the ozone layer might then expose life on land to harmful ultraviolet radiation. Heavy use of nitrogen-rich fertilizers might be increasing these emissions and their damaging consequences. At this point, Crutzen's work attracted little interest.

A few years later, in 1974, Frank Sherwood Rowland and Mario Molina hypothesized that other types of inert gases, the chlorofluorocarbons (CFCs), might also be reaching the stratosphere and destroying ozone. In contrast with nitrous oxide, CFCs were an entirely artificial chemical, produced by industry for use in refrigerators, air conditioners, and even aerosol spray cans. The Rowland–Molina hypothesis provoked an uproar across the industries producing and using CFCs, even while evidence mounted that stratospheric ozone was indeed under threat.

It would take until 1985 before the most serious consequence of ozone degradation would be revealed, in the form of an 'ozone hole' over Antarctica, where a near complete loss of ozone was catalysed by seasonal accumulations of CFCs (Figure 6). The discovery of the ozone hole immediately sparked grave concern not only among scientists, but also among the public and policymakers. Within a few years, an international coordinated effort introduced new policy frameworks to address the ozone hole, beginning with the Montreal Protocol.

The Montreal Protocol and subsequent stricter frameworks would ultimately reduce and eliminate the production and use of CFCs, allowing the ozone layer to recover. Crutzen, Rowland, and Molina shared the 1995 Nobel Prize in Chemistry. And the narrative that human alteration of the atmosphere was producing grave consequences took root, together with the demand for stronger international efforts to detect, understand, and potentially avoid the harmful consequences of human alterations of the Earth system.

The International Geosphere-Biosphere Programme (IGBP)

In 1972, the United Nations held its first conference on the 'human environment' and the UN Environmental Programme (UNEP) was formed. Efforts to stem environmental damage were increasingly supported by an array of government agencies that actively supported research and policy aimed at understanding and engaging with environmental concerns.

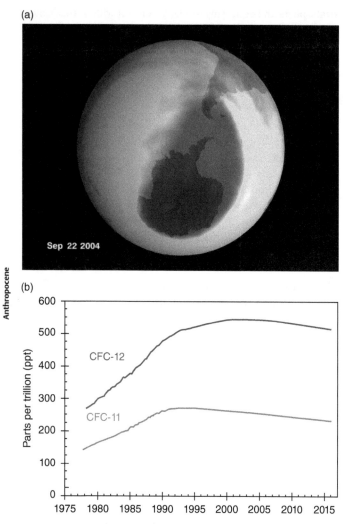

(a)

Sep 22 2004

(b)

CFC-12

CFC-11

Parts per trillion (ppt)

6. The ozone hole over Antarctica and long-term changes in atmospheric CFCs.

As documented in Rachel Carson's influential book *Silent Spring*, artificial chemicals like DDT were harming reproduction in birds and other animals far from sites of application. Farming, grazing, and urban sprawl were rapidly replacing and damaging natural habitats. Acid rain produced by sulphur dioxide (SO_2) emissions from burning coal in one region could move hundreds of miles before falling on another region and even another country, damaging forests and freshwaters there. By the 1980s, it was clear that many environmental problems were increasingly global in scale. A science of global environmental change would be needed to understand and address these problems.

Concern about the ozone hole joined with this broader array of environmental challenges in the call for a more robust science of global environmental change. New international scientific organizations dedicated to the study of global environmental change were commissioned, building on prior global research collaborations like the International Geophysical Year. The first was the World Climate Research Programme in 1979. In 1986, a widely distributed NASA report called for increasing

> scientific understanding of the entire Earth system on a global scale by describing how its component parts and their interactions have evolved, how they function, and how they may expect to continue to evolve on all timescales.

The report included a conceptual model of the Earth system that included the influence of 'human activities' (Figure 7). A new international scientific institution would be needed to support this.

In 1987, Paul Crutzen and Will Steffen were among the first to join the newly formed International Geosphere-Biosphere Programme (IGBP). With the formation of the IGBP, based in Sweden, Earth system science gained the institutional capacity it would need to build a robust interdisciplinary community of scientists dedicated to advancing Earth system science.

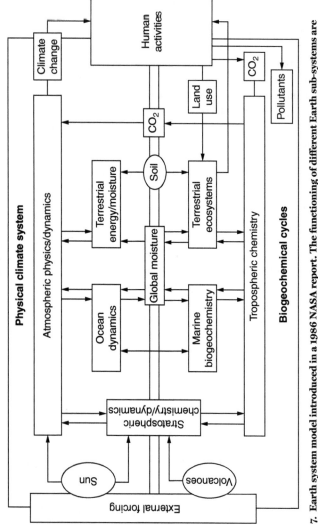

7. Earth system model introduced in a 1986 NASA report. The functioning of different Earth sub-systems are connected with 'human activities' including land use, pollution and emissions of CO_2.

Changing the system

By the mid-1990s, Earth system scientists coordinated by the IGBP and other international scientific institutions had assembled a potent body of evidence demonstrating that humans were dramatically altering Earth's functioning as a system. Not only were human activities filling the atmosphere with carbon dioxide, CFCs, aerosols, and other trace gases, these were threatening Earth's protective ozone layer and driving global changes in climate. The global cycling of the elements, Earth's biogeochemical cycles, were being disrupted, and not only the carbon cycle, but also the cycles of nitrogen and other life-giving elements. Human use of land was reshaping Earth's ecology, eroding productive soils, diverting water to farms and cities, eliminating natural habitats, and driving species extinct at alarming rates.

Observations clearly documented that human activities were changing in tandem with changes in Earth's atmosphere, lithosphere, hydrosphere, biosphere, and climate. By the 1990s, these observations could be taken a step further. Using computer simulations of material and energy exchanges among the spheres, it became possible to confirm that increases in human activities were the cause of, not just correlated with, major long-term changes in the functioning of Earth as a system. This was no minor achievement. Without the ability to conduct experiments on a single Earth, the emergence of this capacity to simulate Earth system processes was a scientific breakthrough. In the words of Hans Joachim Schellnhuber in his groundbreaking 1999 paper in *Nature*, Earth system modelling, combined with global observations from space by remote sensing and global observing networks on ground and sea, represented a 'Second Copernican Revolution' aimed at 'unravelling the mysteries of the Earth's physique, or "Gaia's body"'.

In 2001, the IGBP hosted a pivotal meeting of more than 1,400 members of the scientific, policy, resource management, and media communities in Amsterdam. Focusing on the need to study Earth as a system, the meeting produced the 'Amsterdam Declaration on Global Change', which included the following statements:

> The Earth System behaves as a single, self-regulating system comprised of physical, chemical, biological and human components.

and

> Anthropogenic changes to Earth's land surface, oceans, coasts and atmosphere and to biological diversity, the water cycle and biogeochemical cycles are clearly identifiable beyond natural variability. They are equal to some of the great forces of nature in their extent and impact. Many are accelerating. Global change is real and is happening now.

Earth system science continues to investigate the causes of dynamic changes in Earth's functioning. Perhaps the best studied of these is the claim that humans are now 'overwhelming the great forces of nature'—a claim now supported by conclusive evidence that humans are causing unprecedented changes in Earth's functioning as a system. Moreover, these anthropogenic changes have the potential to produce even more rapid, surprising, and potentially catastrophic consequences as a result of tipping points and other complex feedbacks within the Earth system.

It should come as no surprise that Paul Crutzen's spontaneous demand for a new interval of Earth history came at the IGBP's 2000 meeting in Mexico (Crutzen was vice chair of IGBP at this time). But the large body of evidence from Earth system science is not enough in itself to alter the Geological Timescale—the formal, internationally agreed ordering of 4.6 billion years of Earth history into geological eons, eras, periods, and epochs. To declare a new interval of geologic time, geologists would need to apply their own scientific methods, procedures, and evidence. It would be necessary to show that humans have left a clear, globally identifiable marker in the rocks.

Chapter 3
Geologic time

Eight years after Crutzen's outburst, geologists were ready to act. In their 2008 paper 'Are we now living in the Anthropocene?', Jan Zalasiewicz and colleagues at the Geological Society of London called on geologists to consider the Anthropocene as a new interval of geologic time.

Overwhelming scientific evidence already demonstrated that humans had altered Earth. It might therefore seem strange that this is not yet recognized in the scientific timeline of Earth history, the Geologic Timescale. The reason for this is simple. The keepers of geologic time must build their timelines directly from the physical records of Earth history laid down in rocks by the geological processes that shape our planet.

To understand the challenges of defining the Anthropocene as an interval of geologic time, it is necessary to understand the scientific methods employed by geologists in constructing the Geologic Time Scale. To begin with, one must know that geologic time is deduced from layers, or 'strata', deposited over time, one layer over another, across long periods, producing a layered 'stratigraphic' record. For example, such a record might be preserved in the layers of sediment deposited at the bottom of a lake over many years, all of which might later harden into sedimentary rock.

Geologists who specialize in the study of such stratigraphic records, known as stratigraphers, are the keepers of geologic time. It is the scientific community of stratigraphers who will ultimately decide the fate of the Anthropocene as an interval of Earth history. If we are to understand the case both for and against recognizing the Anthropocene as an interval of geologic time, a basic understanding of the science of stratigraphy and how the Geologic Time Scale is established is essential.

Origins of stratigraphy

Stratigraphy began with the late 17th-century work of Nicholas Steno and his interpretations of the layered structure of sedimentary rocks (Figure 8). His 'Law of superposition', in which newer layers of sedimentary rocks must form on top of older layers, remains the most fundamental concept in stratigraphy. To this he added two principles, holding that no matter what condition or orientation layers of sedimentary rock might now be in, they must originally have formed as horizontal and continuous layers.

8. Beds of turbidite, a sedimentary rock from Cornwall, England, illustrating sedimentary layers.

Steno's principles made it possible for ancient sedimentary rocks to be interpreted as layers of time, no matter how deformed, tilted, eroded, or otherwise jumbled up they had become through various geological processes. Steno and others also recognized that the physical characteristics (mineralogy, texture, colour, etc.) and fossil content within layers could be used to differentiate them from each other and enable correlation of layers across different rock formations even when these were exposed in different places.

A century later, the most transformative work of stratigraphy would come from an English surveyor of mines, canals, trenches, and coal pits. Working literally in the trenches, William Smith became intimately familiar with the varied strata he observed across Great Britain. By correlating fossils and different types of rock across these local observations, he connected them together into stratigraphic layers covering the whole of England, Wales, and Scotland. In a feat still known as 'the map that changed the world', Smith was the first to accurately map the occurrence and exposure of rock layers continuously across large areas. Though his struggles for recognition included a stint in debtors' prison, and much of his recognition was posthumous, Smith is now known as 'the Father of English Geology'. His map still hangs in Burlington House, home of the Geological Society of London.

Stratigraphic science has practical applications from mining to construction. Yet its role in reconstructing geologic time is probably its most far reaching. Geologic time is critical to understanding our planet's early crust, the origins and evolution of life, and even the processes of Earth system change that continue to the present day. And stratigraphers have developed a broad array of tools well beyond the imaginations of Steno or Smith.

Geochronology

The first stratigraphic reconstructions of geologic time were 'relative geochronologies'. By this approach, the relative positioning of rock layers, or stratigraphic 'units', is interpreted as sequences in time, in which lower layers are earlier and upper layers more recent. Separate layers are identified through their physical characteristics, as 'lithostratigraphic units', or based on distinctive fossil biota, as 'biostratigraphic units', or some combination of both. While absolute dates cannot be determined this way, by assembling long sequences of these units, very significant intervals of geologic time could be reconstructed. This allowed evolutionary changes in fossil organisms to be observed, yielding a further stratigraphic principle, of 'fossil succession', in which biota tend to shift together in sequences, providing powerful evidence supporting Darwin's then nascent theory of evolution.

In the mid-18th century, Giovanni Arduino and fellow stratigraphers made the first attempts to assemble a continuous timeline encompassing all of Earth's history. In this first calendar of geologic time, four different time intervals, or 'orders', were identified with four different types of rocks, and these were labelled in sequence from Primary to Quaternary. Estimates were also made of the duration of different time intervals based on their thickness and rates of formation. These rates were estimated based on the pace of chemical and physical rock breakdown (weathering), erosion, sedimentation, and the compression and cementing of sediments into solid rocks (lithification).

At the turn of the 20th century, a new technique revolutionized stratigraphy. Radiometric dating, of which 'carbon dating' is a popular example, enabled absolute dates to be assigned to stratigraphic units for the first time, yielding 'geochronologic units' with known dates of formation.

Radiometric dating relies on the principle that some elements have isotopes (variants with different numbers of neutrons) that are radioactive, with different rates of radioactive decay, or 'half lives', as they decay into other elements and isotopes. For example, the most common isotope of carbon is carbon-12 (or ^{12}C), with 6 protons + 6 neutrons. Carbon-12 is stable and does not decay. But carbon also occurs naturally in the form of carbon-14 (or ^{14}C), with 8 neutrons. Carbon-14 is radioactive and decays to half its amount in a period of 5,730 years (its half life). By measuring the relative quantities of different isotopes in a sample of rock or other material, absolute ages can be computed from the amount of each isotope remaining and their relative half lives. While the half life of carbon-14 is brief, at 5,730 years, limiting its use to dating carbon-rich materials no more than 40,000 years old, some elements have isotopes with half lives of hundreds of millions of years, such as Uranium-235 (half life about 700 million years), which can be used to date chronostratigraphic units that are more than one billion years old.

In 1913, radiometric dating was used to determine the age of a rock sample at 1.6 billion years. And that was only the beginning. With the rise of absolute dating and absolute geochronology, geologic time scales could be constructed in linked units of time and rock. Stratigraphy would also gain additional tools, including chemostratigraphy, enabling stratigraphic units to be identified and correlated based on their detailed chemical and isotopic composition, and magnetostratigraphy, in which the age of units could be determined relative to historical reconstructions of changes in the magnetic polarity of Earth. Ultimately, by combining radiometric dating together with this expanded stratigraphic toolkit, more than 4 billion years of Earth history would be reconstructed in magnificent detail.

The Geologic Time Scale

The Geologic Time Scale (GTS) brings together the work of generations of stratigraphers within a single standardized geochronology of Earth history (Figure 9). Such a massive coordination of scientific work was accelerated by the formation of the International Commission on Stratigraphy (ICS) in 1974, as a working committee within the International Union of Geological Sciences, the international coordinating body for geologic science.

From the beginning, the main focus of the ICS has been to organize Earth's history into the GTS based on a standardized hierarchy of chronostratigraphic units. Eons are its largest unit, subdivided into successively shorter units of eras, periods, epochs, and ages. That ICS has been able to produce such an orderly structure of geologic time is all the more remarkable considering that it represents the integration of stratigraphic work conducted at sites around the world starting in the late 18th century. For example, the Jurassic Period was established by Leopold von Buch in 1839, based on Alexander von Humboldt's observations in 1795 of rock formations in the Jura Mountains of Switzerland. Since its first publication in 1982, the GTS has been regularly revised and updated, as new palaeontological evidence has come to light, and dating techniques have improved.

The GTS divides 4.55 billion years of Earth history into chronostratigraphic units that capture many of the key events in Earth history, but not others. Of the five mass extinction events commonly recognized in Earth history, in which exceptionally large numbers of species were lost within short intervals of time, four coincide with period boundaries. The most dramatic, which almost destroyed life altogether, occurred at the end of the Permian Period, currently dated at 252 million years ago, while the most famous, the mass extinction of the non-avian dinosaurs and marine reptiles, occurred at the Cretaceous–Palaeogene boundary (formerly known as the K-T boundary), 66 million years ago. Clear fossil records of multicellular animals and intense burrowing appear at the lower boundary of the Cambrian Period, 541 million years ago, which also begins our current Eon, the Phanerozoic (the 'period of visible life').

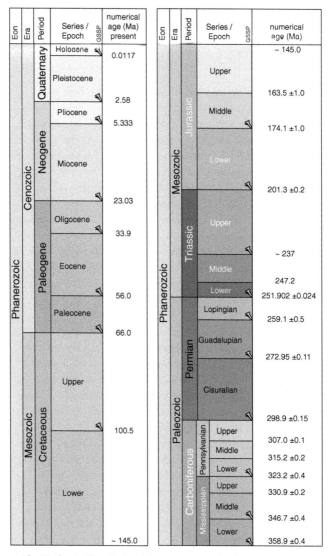

9. The Geologic Time Scale of the International Commission for Stratigraphy (ICS), illustrating Eons, Eras, Periods and Epochs (based on GTS 2017; Ma = millions of years before present).

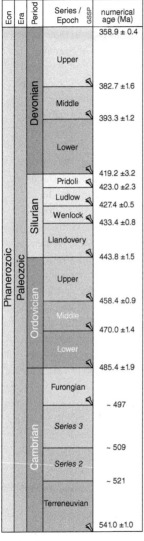

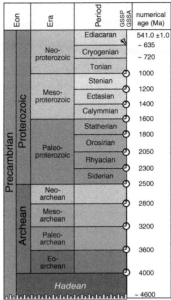

Geologic time

Units of all ranks are in the process of being defined by Global Boundary Stratotype Section and Points (GSSP) for their lower boundaries, including those of the Archean and Proterozoic, long defined by Global Standard Stratigraphic Ages (GSSA). Charts and detailed information on ratified GSSPs are available at the website http://www.stratigraphy.org. The URL to this chart is found below.

Numerical ages are subject to revision and do not define units in the Phanerozoic and the Ediacaran; only GSSPs do. For boundaries in the Phanerozoic without ratified GSSPs or without constrained numerical ages, an approximate numerical age (~) is provided.

Numerical ages for all systems except Lower Pleistocene, Upper Paleogene, Cretaceous, Triassic, Permian and Precambrian are taken from 'A Geologic Time Scale 2012' by Gradstein et al. (2012); those for the Lower Pleistocene, Upper Paleogene, Cretaceous, Triassic, Permian and Precambrian were provided by the relevant ICS subcommissions.

Colouring follows the Commission for the
Geological Map of the World (http://www.ccgm.org)

9. Continued.

Yet the GTS has no geologic boundary marking the origins of life (occurring sometime in the early Archean), the first photosynthetic oxygen-producing organisms (within the Palaeoproterozoic), the first multicellular animals, or the emergence of plants on land (both within the late Neoproterozoic), or even the first animals on land (probably in the Silurian). The reasons are entirely pragmatic. Without identifiable stratigraphic markers, even the most important milestones in Earth history cannot be included in the Geologic Time Scale.

The events in Earth history that are marked in the GTS are only those that have left clear and recognizable global stratigraphic signatures, like the iridium-enriched layer deposited by the massive meteorite impact hypothesized to have caused the dinosaur extinction. Earth's oxygenation by photosynthetic organisms, certainly one of the greatest Earth system changes ever, was simply too gradual to leave an identifiable stratigraphic marker. In contrast, the presence of readily identified animal fossils—the main basis for biostratigraphy—marks a profound division in the GTS, the divide between our present Eon and the Precambrian, a general term referring to all 4.06 billion years of Earth's prior history. This is true even though the rise of multicellular animals began tens of millions of years before this time; but these first species had soft bodies that left few clear fossils.

The methods of stratigraphy are clear. Transformative changes in the Earth system are not enough. To appear in the Geologic Time Scale, an event must leave the right kind of stratigraphic evidence.

Golden spikes

The Geologic Time Scale is divided into time intervals through the identification of stratigraphic boundaries, with the lower boundary of one interval serving as the upper boundary for the one before it (these are known as 'boundary stratotypes'). Since 1977, these boundaries have been defined using markers identified and dated within stratigraphic sequences, usually biostratigraphic signatures such as the first appearance of a fossil organism. These defined and dated markers, known informally as 'golden spikes', identify a 'specific point in a specific sequence of rock strata', and are known formally as Global Boundary Stratotype Sections and Points (GSSP).

The effort to mark all time boundaries in the GTS with GSSPs remains a work in progress. Owing to the paucity of fossil evidence, Precambrian boundaries are mostly marked by chronologic times, or Global Standard Stratigraphic Ages (GSSA), instead of GSSPs. Nevertheless, the ultimate goal is to mark all time intervals in the GTS with peer-reviewed and published GSSPs.

GSSPs are much more than just dated points in rocks. After marking a specific point within a specific stratigraphic sequence, each GSSP is formally registered and preserved in an accessible place, enabling future observations. For example, the GSSP marking the Precambrian–Cambrian boundary is identified at the first appearance of the characteristic fossil traces, named *Treptichnus pedum*, of a burrowing species in a rock sequence located in a nature preserve in Fortune Head, Newfoundland (the 'Fortunian GSSP'; Figure 10).

While some GSSPs are marked at their site of origin using an actual, metal, 'golden spike', this is not required. What is required is that each GSSP should represent the 'best possible' record of both the boundary marker and the stratigraphic sequence above and below it. One major concern is that the boundary level identified by the GSSP should be 'isochronous', representing a chronostratigraphic unit that can be identified at multiple sites around the world at the same time, rather than a 'diachronous' unit varying in age from place to place. To avoid selecting diachronous markers, multiple stratigraphic sequences must be observed in a variety of different locations around the world and

10. Example of a Global Boundary Stratotype Section and Point (GSSP), or 'golden spike', marking the base of the Ediacaran Period. Located in Ediacara, South Australia.

examined together as a 'global synthesis'. Moreover, the ideal GSSP should also be datable using radiometric or other reliable techniques, and should include multiple distinct markers, both biostratigraphic and others (magnetostratigraphic, chemostratigraphic), that can be correlated in time across stratigraphic sequences around the world. As we shall see, the need to reject diachronous markers—and therefore, diachronous environmental processes—has emerged as a point of debate in discussions concerning defining a GSSP for the Anthropocene.

Taken together, these rigorous requirements are often difficult to fulfil. Pragmatic solutions to numerous stratigraphic issues are normally required. The completion of a single GSSP proposal can demand years of meticulous research. If conditions are met, a working group's GSSP proposal is submitted for peer review. Following a successful vote on the GSSP within the working group, parent subcommission, the ICS itself, and finally the Executive Committee of the International Union of Geological Sciences, a GSSP may be ratified and registered within the GTS. By this formal, international, scientific institutional procedure, the history of Earth is connected with the physical records of Earth history written in rock. This is the process that might allow an Anthropocene Epoch to be marked geologically as the newest time interval in the GTS.

The Quaternary

Earth's most recent period, beginning 2.6 million years ago, is the Quaternary, and it is therefore the interval within which the Anthropocene Epoch would most likely be defined. The Quaternary exemplifies the challenges and opportunities of geologic time, with roots extending deep into the early days of stratigraphy. As the only remaining 'order' of Arduino's 1759 four 'order' Earth calendar that is still used in the GTS, even the Quaternary was omitted from the GTS for five years, returning in 2009. Moreover, as our species evolved entirely within the Quaternary, this has inspired alternative names—for example the term 'Anthropogene' was preferred by Soviet geologists in the 1980s.

The Quaternary represents a relatively cold interval of Earth history, also known as the 'current ice age'. It is distinguished from the Neogene Period before it by more intense glacial/interglacial cycles and more extensive sheets of continental ice during its cold, glacial intervals. Covering Earth's most recent 2.6 million years, the stratigraphic records of the Quaternary are generally more abundant, more accessible, and more detailed than those of earlier times. For this reason, geologists have also been able to reconstruct a variety of continuous records of Earth system changes in addition to the discrete boundaries used in the GTS. For example, the glacial/interglacial cycles of the Quaternary have been reconstructed in detail by measuring changes in oxygen isotopes in cores of deep ocean sediment (Figure 11).

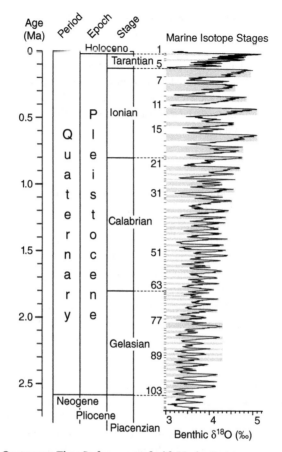

11. Quaternary Time Scale compared with Marine Isotope Stages (MIS). MIS are delineated by major global shifts in temperature indicated by changes in Oxygen isotopes ($\delta^{18}O$). Odd numbered MIS are warm periods; MIS 1 overlaps with the Holocene interglacial.

The lighter isotope of oxygen, oxygen-16, evaporates more easily from the sea, leaving behind water enriched in the heavier oxygen-18. And when ice accumulates on land during glacial intervals, oxygen-16 becomes locked up in the ice, leading to seas and sediments further enriched in oxygen-18. By measuring ratios of oxygen-18/oxygen-16 in sediments over time, Earth's warming and cooling can be inferred with considerable precision. Working back from the present, each 'stage' of Earth's warm interglacial and cool glacial cycle has been numbered in a widely used system of 'Marine isotope stages' (MIS), starting with MIS 1 for the current warm interglacial (the Holocene Epoch), preceded by MIS 2, as the 'last glacial maximum', and so on, all the way back into the Pliocene. These cyclic stages have also been linked with long-term changes in atmospheric carbon dioxide and other trace gases measured in ice cores, and are commonly applied in reconstructing ecological and archaeological timelines.

Most of the Quaternary is comprised of the many glacial/interglacial cycles of the Pleistocene Epoch ('most recent'). The Holocene Epoch ('wholly recent') begins only 11,700 years before present, marking the shift to our current warm interglacial interval. The Holocene is unique in having its lower boundary marked by a GSSP within a core of solid ice extracted in Greenland (Figure 12). Ice is a form of rock (a mineral solid), and the Greenland ice sheet

Geologic time

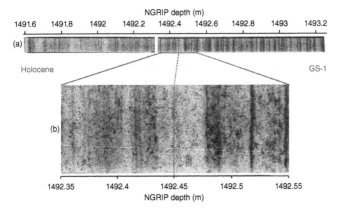

12. The Holocene GSSP. The lower boundary of the Holocene is marked at a depth of 1492.45 metres in an ice core extracted from the Greenland ice sheet.

47

is formed from annual layers of snow that have been compacted over time. The stratigraphic record provided in continental ice is more precisely defined and consistent than that of ocean sediments, which can become mixed and disordered by animal movements ('bioturbation') and other processes. For this reason, even though the Holocene GSSP aims to mark the beginning of the current interglacial, as does the sedimentary record marking MIS 1, the ice record provides a more precise start time than would be possible using marine sediments (11,700 +/– 100 years before present).

The Quaternary has long served as a test bed for advancing the techniques and theory of geologic time, including methods for counting back from the present time, the use of ice cores, and the reconstruction of detailed and continuous timelines using isotopic methods, radiometric dating, and chemostratigraphy. To define the Anthropocene as a geological unit within the GTS, even more novel stratigraphic approaches might be needed.

The Anthropocene Working Group

In 2009, the ICS Subcommission on Quaternary Stratigraphy recommended Jan Zalasiewicz, a professor at Leicester University, UK, and an expert on biostratigraphy, to form an Anthropocene Working Group (AWG). The new group would have a single task: to examine the case for recognizing a new interval of geologic time based on 'the wide-ranging effects of anthropogenic influence on stratigraphically significant parameters'. In other words, the AWG would examine the case for subdividing the Quaternary Period of the GTS by identifying the lower boundary of a potential Anthropocene Epoch, ideally with a new GSSP. With Zalasiewicz as chair, the AWG was formed within the year with sixteen members, about half stratigraphers and the rest a mix of environmental scientists with expertise in anthropogenic global change, including Paul Crutzen, Will Steffen, and myself, and even a lawyer, Davor Vidas, an expert on the Law of the Sea.

Working part time, without funding, the AWG began slowly. Unlike prior intervals of geologic time, the stratigraphic basis for recognizing the Anthropocene would need to 'critically compare the current degree and rate of environmental change caused by anthropogenic processes with the environmental perturbations of the geological past'. This was a novel requirement for stratigraphic work. Also unlike earlier intervals, data on recent Earth changes capable of leaving stratigraphic records, both natural and anthropogenic, were hyper-abundant, ranging from changes in global climate and atmospheric composition, to ocean chemistry, biodiversity loss, environmental pollution, increases in soil erosion, and massive alteration of landscapes across entire regions. Disentangling this abundance of data made the job harder, not easier. There were also questions about the utility to geologic science of formalizing the Anthropocene—still a matter of debate. Fortunately, the need for a formal GSSP proposal was still years away.

Zalasiewicz's question, 'are we in the Anthropocene?', was perhaps not the hardest one. Scientific consensus already recognized that human transformation of Earth was well under way and leaving abundant stratigraphic evidence. In practice, therefore, the main question facing the AWG is not whether, but when, and on what basis, the Anthropocene might be recognized within the GTS. An Anthropocene GSSP might be identified within layers of sediment or ice or other materials, or even defined chronologically, by a GSSA. The possibility of an age or even period was also on the table, though an epoch defined by a golden spike was clearly the favoured approach (Figure 13). And multiple possible markers were already under consideration.

Paul Crutzen had tied the Anthropocene to the late 18th century and the Industrial Revolution, with combustion of fossil fuels causing an initial uptick in atmospheric carbon dioxide concentrations above those typical of the Holocene. Will Steffen, building on this earlier proposal, came to argue for the

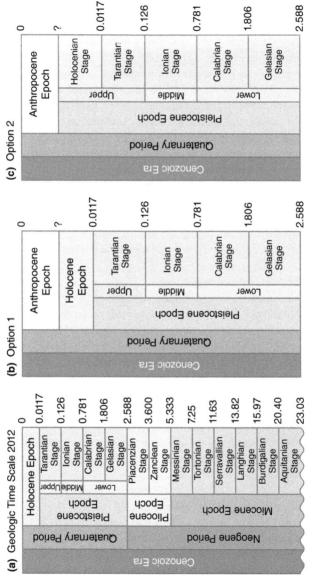

13. **Potential revisions of the Quaternary to include the Anthropocene.** (a) **Existing Geological Time Scale (GTS 2012;** numbers at right are millions of years). (b) **Option 1:** Holocene epoch ends, followed by an Anthropocene epoch. (c) **Option 2:** Holocene replaced by an Anthropocene epoch, Holocene reduced to a Stage within the Pleistocene (this option is not currently under consideration by the AWG).

mid-20th century as the main point of onset for the Anthropocene, marked by the 'Great Acceleration' in human activities around that time. And geologist Bill Ruddiman even suggested that the Anthropocene might be recognized thousands of years before the Industrial Revolution as the result of widespread land clearing for agriculture, causing releases of carbon dioxide and methane, and potentially global climate change. All of these proposals offered the prospect for an Anthropocene GSSP. Yet from a pragmatic, stratigraphic, point of view, only one proposal presented a relatively straightforward, unambiguous, basis for a global, isochronous, stratigraphic marker: the spread of radioactive fallout from nuclear weapons testing, beginning with the Trinity Test of 1945.

Chapter 4
The Great Acceleration

For Will Steffen and others at IGBP, the claim that humans
were transforming Earth's functioning as a system was not new.
Evidence for this had been accumulating for decades. Indeed, this
view was already mainstream among environmental scientists.
The challenge for Steffen and his team, tasked in 1999 with
reviewing a decade of Earth system research for the IGBP, was
the opposite of scarcity. The challenge was how to integrate
thousands of articles and reports into a coherent overview
of global environmental change from an Earth system
science perspective.

Inspired by Crutzen's vision of the Anthropocene, the team
focused on human-driven changes to the Earth system brought
on by the Industrial Revolution. They amassed records of human
activities and environmental changes starting in 1750—before
James Watt advanced the steam engine—and plotted their
dynamics through to the year 2000. Though the evidence clearly
confirmed human transformation of the Earth system, what they
found surprised them. Published in 2004 as the now classic IGBP
report *Global Change and the Earth System: A Planet Under
Pressure*, their work did not reveal a continuous rise in Earth
transformation as the Industrial Revolution gathered steam and
spread across the world. Rather, the data showed a dramatic jump
in the rate of human and environmental changes starting in the

middle of the 20th century. Depicted in two panels of twelve charts, their work exposed a striking inflection point around 1950 in virtually all of the human activities and Earth system changes they examined, after which rates of change become far steeper and in some cases almost exponential (Figures 14 and 15).

The message from Earth system science was clear. Starting in the 1950s, humans began shifting Earth's functioning as a system into a new and unprecedented state. As stated in their report:

> The last 50 years have without doubt seen the most rapid
> transformation of the human relationship with the natural world in
> the history of humankind.... The magnitude, spatial scale, and pace
> of human-induced change are unprecedented in human history
> and perhaps in the history of the Earth; the Earth System is now
> operating in a 'no-analogue state'.

In 2005, in analogy to Karl Polanyi's *The Great Transformation*, 'The Great Acceleration' was coined and began making the rounds in scientific circles as the common term describing the dramatic mid-20th-century step-change in anthropogenic global environmental change. The charts demonstrating these changes would soon come to symbolize the Anthropocene both inside and outside the scientific community. From an Earth system perspective, the Anthropocene began in the middle of the 20th century.

Planet Under Pressure

By presenting a broad suite of human and environmental changes as the basis for understanding Earth system change, *Planet Under Pressure* not only made the case for a new interval of Earth history, but also for the need to understand anthropogenic global environmental change as a complex, multi-causal, system-level set of processes affecting the entire Earth system. Humans were doing a lot more than just changing Earth's atmosphere and

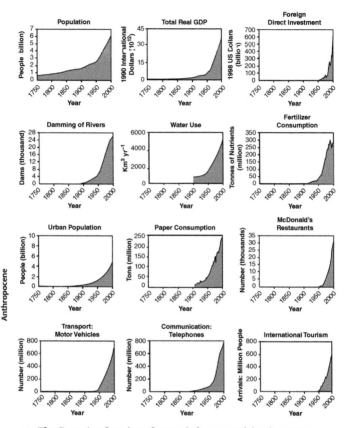

14. The Great Acceleration: changes in human activity since 1750.

climate, they were also causing global changes in biodiversity, polluting the oceans with fertilizer runoff from agriculture, altering the flows of rivers to the sea, and transforming natural habitats around the world. Human influences in the global environment could not be reduced simply to fossil fuel combustion or the production of industrial chemicals. Population growth, the 'domestication' of land for agriculture, economic

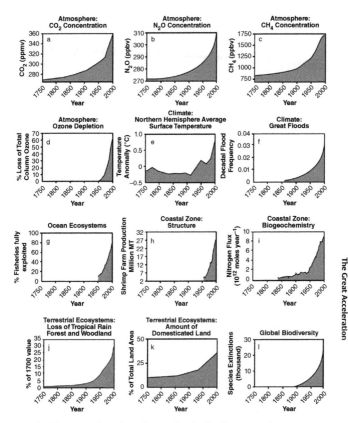

15. **The Great Acceleration: changes in the Earth system since 1750.**

development, and even foreign direct investment were all part of the mix of human driving forces that were altering Earth's functioning as a system. Anthropogenic global environmental change was a multidimensional process. Moreover, interactions among its many dimensions and cascading effects from the local to the regional and ultimately to the global might lead to unanticipated consequences for the Earth system as a whole. And a key principle was put forward.

To recognize the global significance of anthropogenic changes from an Earth system perspective—the basis for observing a planetary transition to an 'Anthropocene Era'—it would be necessary to demonstrate that humans had caused Earth system processes to change beyond their 'range of natural variability', into a 'no-analogue state'. This 'natural range' in an Earth system attribute, such as temperature or atmospheric carbon dioxide, would first need to be characterized in terms of long-term patterns of variation observed over intervals of Earth history—half a million years or longer in one estimation. Then, evidence would need to demonstrate that human activities had forced this attribute outside this natural range. Local or even regional environmental changes would not be enough.

Domesticating land

Humans first began using land for agriculture and settlements more than 10,000 years ago. Yet the scale, extent, intensity, and rates of land conversion for human use all increased dramatically in the industrial age. Global estimates of contemporary land use vary, but generally indicate that 40 per cent to 50 per cent of Earth's ice-free terrestrial surface is now in use for agriculture, forestry, and human settlements. Approximately 11 per cent of Earth's land is cultivated for crops, 25 per cent is used for pastures and livestock grazing, and 1 to 3 per cent forms urban and other settlements and infrastructures. Woodlands managed or planted to produce timber, fuel, paper, rubber, and other products occupy another 2 to 10 per cent of Earth's land. Of land remaining without these direct intensive uses, at least half or more is altered by local fuel gathering, hunting, foraging, pollution, and other local human influences. As a result, three-quarters of the terrestrial biosphere has been transformed directly and indirectly by human use of land. Less than one-quarter remains free of direct human impacts, and mostly in the less productive, colder, and drier regions of the terrestrial biosphere, though some also remain in the Tropics, where endemic diseases and other obstacles limit human settlement.

The environmental consequences of human use of land range from greenhouse gas emissions to environmental pollution, soil erosion, habitat loss, species extinctions, and species introductions. Yet there is perhaps no greater single environmental transformation than the cultivation of crops, beginning with the clearing of land and tillage. Vegetation is removed, usually by burning, emitting carbon dioxide. Soils are uncovered, leading to erosion and loss. Disturbance, tillage, and the draining of wetlands cause abundant soil organic matter to decompose, releasing yet more carbon dioxide. Flooding soils to produce rice releases large amounts of methane gas (CH_4), each molecule having more than ten times the greenhouse warming potential of carbon dioxide (though it spends less time in the atmosphere than CO_2). Use of nitrogen-rich fertilizers (both manures and synthetic fertilizers) releases nitrous oxide (N_2O), an even more potent greenhouse gas with more than 100 times the warming potential per molecule of carbon dioxide, and very stable.

Use of pesticides and herbicides harms species both on and off agricultural fields, joining excess nutrients from fertilizers in polluting ponds, lakes, streams, rivers, and coastal areas downstream. Livestock agriculture displaces native herbivores through direct competition and by the act of controlling predators and competitors. In many cases, land is also cleared to improve vegetation productivity for livestock, while intensive large-scale livestock systems ('factory farms') produce methane and other greenhouse gas emissions from manures, together with environmental pollution similar to but often more concentrated and hazardous than that from cropped fields. Domesticated chickens are now Earth's most abundant bird and cattle biomass alone exceeds that of all other living vertebrate animals combined—including humans.

The global impacts of agricultural and urban areas, beyond their greenhouse gas emissions, effects on soils, and pollution of water, soils, and air, also transform, replace, and displace native habitats and native species. While low intensity land uses, such as grazing or forestry, may have relatively small effects on many species, the effect on species sensitive to human influences may represent a complete loss of habitat, denying these species the resources needed to sustain viable populations. At the same time, humans have transported and introduced species around the world, both intentionally as crops, ornamentals, pets, and to control other species, and also unintentionally by facilitating their movements as hitchhikers on human transport networks. While most introduced species either fail to reproduce or establish only minor populations, some invade rapidly and establish large populations across landscapes, displacing existing species, especially in landscapes already altered by human activities. Combined together, habitat loss, hunting, foraging, pollution, species invasions, and other human pressures are increasingly threatening populations of vulnerable plant and animal species with extinction, leading to rapid global losses of biodiversity. While land use for agriculture and settlements transformed significant areas around the world long before 1950, human population growth and the richer diets supported by industrial economic development caused both a rapid global expansion of land use and broad increases in land use intensity, including a major rise in the use of irrigation and agricultural chemicals.

Hydrosphere

Human engineering of the hydrosphere began more than 5,000 years ago with the construction of irrigation ditches, canals, dams, reservoirs, the diversion of rivers and streams, the digging of wells to extract groundwater, and other water control systems designed to support agricultural production and human settlements. Irrigation of agricultural fields remains the dominant use of water redirected by humans, representing 60 to 75 per cent of approximately 5,000 km^3 flowing through human-engineered systems annually, as of 2000. Annually, about 40,000 km^3 of freshwater flows into the sea as runoff across the continents, but less than one-third of this flow is accessible to human societies owing to seasonal and geographic variations in its distribution. As a result, human societies are already using nearly half of the renewable freshwater flows available to them.

The construction of dams, reservoirs, rice paddies, and other impoundments has increased water storage on land and delayed water flows to the sea. Wetland drainage for agriculture and development has done the opposite. Water release from vegetation to the atmosphere has been altered by land clearing for agriculture, forestry, and by built infrastructure, and the balance of runoff versus retention of water in soils and groundwater has also been altered in opposing ways by the covering and tilling of soils. Moreover, these widespread hydrological changes have altered water availability not only for humans but also for fish and other aquatic and terrestrial species that depend on unaltered flows of water, in terms of both quantity and seasonal and spatial availability. Dams and other water diversions also restrict the movement of sediments and pollution to the sea, causing harmful accumulations upstream and reducing sediment deposits in coastal areas, leading to soil subsidence and exposing coastal infrastructure to rising seas.

Freshwater appropriations have long increased with population and food demand and therefore accelerated after 1950. Over the second half of the 20th century, rapid and unsustainable extraction of groundwater and massive dam projects for both agriculture and power generation helped accelerate the full range of human hydrological modifications across the continents, though large dam projects have declined in recent years. Freshwater availability has also been reduced by surface and groundwater pollution with industrial chemicals and excess nutrients. As a result, since the 1950s, the terrestrial hydrosphere—Earth's freshwater systems—has been transformed profoundly by human activities and limits to freshwater availability have become a matter of serious global concern.

Biosphere

Humans began transforming the biosphere long before agriculture. Even before the Holocene, hunting and foraging pressures on terrestrial, freshwater, and marine species caused local populations to decline and caused a number of global species extinctions. As human populations spread and grew, hunting and foraging pressures generally increased, though the rise of agriculture displaced these pressures to some degree. With the spread of agriculture, terrestrial species retreated to shrinking habitats, and habitat loss became the primary driver of population declines and extinctions for non-prey species. The story for aquatic species is different.

Increasing societal demands for seafood, including freshwater species, have continued to put intense harvest pressures on wild species dwelling in freshwater, coastal, and marine environments. Outside of a few freshwater and coastal habitats, traditional hunting and foraging pressures were generally insufficient to cause major population declines and extinctions and the open ocean remained only lightly influenced. All of that changed with industrial scale fishing, as fleets of 'factory ships' expanded across the oceans. As populations and demands for seafood grew after 1950, fishing also grew in scale and intensity, including the use of massive nets dragged across the seafloor. At the same time, coastal habitats were increasingly transformed by agricultural runoff and the construction of urban areas and other infrastructure, including the removal of mangroves and other wetland systems, altering areas key to reproduction for many species.

Adding to habitat loss and direct exploitation, extinction rates and the functioning of the biosphere as a whole have also been affected by water pollution and anthropogenic changes in the biogeochemical cycles of nitrogen and phosphorus. Industrial toxic pollutants that spread through water, from lead to DDT, have harmed species both directly and by the accumulation of toxins up the food chain, as contaminated organisms are consumed in large quantities by predators. Perhaps surprisingly, excess nutrients, in the form of reactive nitrogen and phosphorus, can produce similar, and, in some cases, even more extreme, effects on aquatic species and habitats than toxic pollutants.

Phosphorus, like nitrogen, is a limiting nutrient for crop growth. While needed in smaller amounts, reactive phosphorus (in various forms of phosphates; PO_4), has been mined, processed, and applied as fertilizer in accelerating amounts since the 1950s. Excess phosphorus, mostly bound to soil particles, may be washed into freshwater streams, rivers, ponds, or lakes, enriching them with nutrients, a process called eutrophication. This stimulates the growth of microscopic plants, algae and phytoplankton, and photosynthetic bacteria (cyanobacteria). The resulting blooms of algae and cyanobacteria block light from the waters below, thus inhibiting the growth of seagrasses and other plants that sustain critical seafloor and lakebed habitats. Eutrophied water bodies—often with pungent greenish waters (cyanobacterial blooms smell bad)—are commonly observed in agricultural and coastal regions and also wherever urban sewage or livestock manures rich in phosphorus enter water untreated. Reactive nitrogen produces similar effects in coastal areas, where rivers carry nitrogen to the sea. This is because nitrogen, while rarely a limiting nutrient in freshwater, is extremely scarce in seawater. So the most extreme eutrophication events occur in coastal oceans. 'Dead zones' in coastal areas are produced wherever excess nitrogen yields huge algae blooms, which sink, decompose, and in the process use up so much oxygen that sea creatures are unable to breathe. Incidences of toxic algal blooms, such as red tides, have also greatly increased since the 1950s in both freshwater and coastal areas.

Nearly every form of human-induced pressure on species and ecological processes across the biosphere has increased dramatically since the 1950s. Human use of land has displaced and polluted natural habitats at the same time that wild species have been increasingly exploited. For all of these reasons together, contemporary rates of human-driven extinctions on land and sea, especially of animal species, have increased dramatically since the 1950s, and are now far higher than extinction rates across most of Earth's history.

Nitrogen

Changes in Earth's carbon cycle are often presented as the prime evidence of a human-altered planet. Yet there are many reasons why anthropogenic changes in Earth's global biogeochemical cycle of nitrogen are far more significant. Fossil fuels are part of this, but the vast majority of this unrivalled anthropogenic global change has been the result of a single industrial process that both generates more reactive nitrogen than any natural process and has made it possible to sustain the unprecedented growth of human populations over the past half century.

Nitrogen is a basic component of protein and therefore an essential nutrient required by all living organisms, including the crop plants that provide our food. Without industrially synthesized nitrogen fertilizer and its ability to boost crop yields on limited land, food production could never have kept pace with the demands of more than 4 billion humans after 1970, let alone the needs of 7 billion today or the 11 billion expected in 2100.

Nitrogen, in the form of highly stable and unreactive N_2 gas, is the most abundant element in Earth's atmosphere (78 per cent by volume). Yet, perhaps surprisingly, it is also the most common limiting nutrient for plant growth on land and in the sea. This is because plants (and most bacteria) can only take up and utilize reactive, 'available', forms of nitrogen: as ammonium (NH_4^+) and nitrate (NO_3^-) ions. The process of converting stable N_2 into available nitrogen requires huge amounts of energy. Only a few bacterial species have evolved the specialized high-energy metabolism needed to 'fix' nitrogen, by cracking N_2 to produce ammonium, though many bacteria can easily convert ammonium to nitrate. Worse still, reactive nitrogen is easily lost from soils and water by leaching and runoff, and is lost back to the atmosphere when microbes convert it back into stable N_2 or N_2O gas ('denitrification'), and when biomass is harvested or burned, or dead organisms sink to the seafloor out of reach of photosynthetic plants at the sunlit surface. Until 1910 and the work of Fritz Haber and Carl Bosch, nitrogen was always in short supply, and crop yields were low. The only way to obtain reactive nitrogen for fertilizer was to mine it (deposits of fossilized bird guano), to harvest manures and biomass, or to grow legumes—plants supporting a symbiosis with bacteria capable of fixing nitrogen.

Haber won the Nobel Prize in 1918 and the Haber–Bosch process transformed Earth's nitrogen cycle. By combining large amounts of energy and carbon (usually methane) with N_2 gas, their process fixed nitrogen into ammonium that could be used for fertilizer and other industrial processes, including bombs. Synthetic nitrogen fertilizers dramatically increased crop yields, often doubling them or more, especially when modern crop varieties were bred to take advantage of them—the basis for the 'Green Revolution' in agriculture that spread across the world beginning in the 1950s. With far higher yields, crop production increased without the need for an equivalent increase in land given over to cultivation; most of the 20th century's expansion of agricultural land use has supported livestock production. At the same time, nitrogen fertilizers, especially when over-applied, have polluted ground and surface waters with nitrates, causing health risks, and saturated coastal ecosystems with nitrogen, producing algal blooms and dead zones. Emissions of nitrous oxide from fertilized fields are now an increasingly significant source of greenhouse gases in Earth's atmosphere. In addition to artificial nitrogen fixation, the combustion of coal, petroleum, and biomass also releases acidic forms of nitrogen gases, the nitric oxides (NO and NO_2), that together with sulphur oxides produce 'acid rain', which caused widespread environmental damage in the 1980s and 1990s before these acidic emissions from coal-fired power plants and unregulated vehicle engines were regulated and brought under control.

Artificial nitrogen fixation for fertilizer, nitrogen fixing crops, and fossil fuel combustion together now fix significantly more reactive nitrogen than all natural processes in the terrestrial biosphere combined (Figure 16). By comparison, if anthropogenic carbon dioxide emissions were to exceed all natural emissions from land, they would need to increase by a factor of 10. Anthropogenic transformation of the global biogeochemical cycle of nitrogen is one of the most striking examples of human alteration of Earth's functioning as a system (Figure 17). Before the 20th century, the industrial process behind this profound Earth system

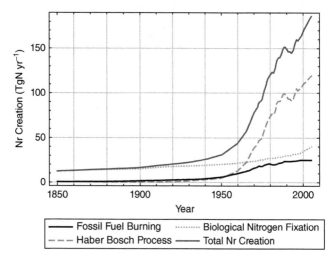

16. **Global changes in reactive nitrogen (Nr) since 1850** (Tg = Teragrams = 10^{12} grams).

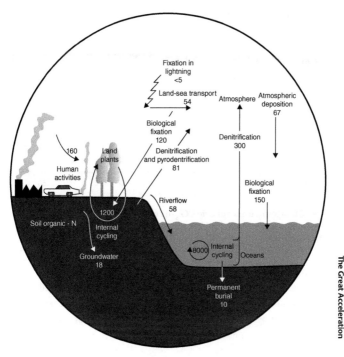

17. The global nitrogen cycle (numbers are Tg N).

transformation did not even exist. From the 1950s on, artificial nitrogen fixation accelerated, helping human transformation of Earth to reach unprecedented levels.

Atmosphere and climate

Anthropogenic greenhouse gas emissions and their effects on Earth's atmosphere and climate are among the most compelling evidence of an accelerating planetary transition caused by humans. Starting with Keeling's Curve, global atmospheric changes have been monitored and also reconstructed in detail into the deep past. Carbon dioxide, methane, and nitrous oxide all

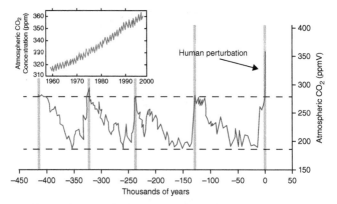

18. **Changes in atmospheric CO$_2$ over the past 450,000 years, illustrating extremely rapid recent increase above past levels. Inset is Keeling Curve detailing changes observed since 1960.**

increased steeply across the last century, rising to levels unprecedented across the entire Holocene. CFCs are the only exception, booming from the 1950s to the 1990s when they were phased out by international agreements to protect stratospheric ozone, which is now recovering.

Atmospheric carbon dioxide has always varied substantially over time. Nevertheless, anthropogenic changes in carbon dioxide concentrations are well beyond their geologically recent natural range of variability (Figure 18). Today's levels of carbon dioxide (>400 ppm) are almost certainly higher than they have been at any time for the past 4 million years or even longer. Rates of atmospheric temperature change are also exceptionally rapid, and are accelerating together with rates of anthropogenic carbon dioxide emissions since 1950 (Figure 19).

Rises in Earth's mean surface temperature closely track anthropogenic changes in atmospheric carbon dioxide and other greenhouse gases (Figure 20). This close correlation only adds to evidence from competing Earth system simulation models that

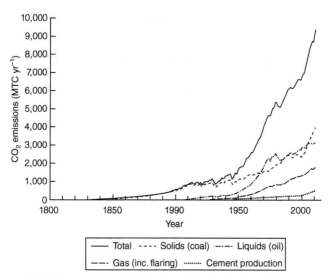

19. **Global changes in anthropogenic carbon dioxide emissions, 1800 to 2000 from different sources, including fossil fuels and cement production.**

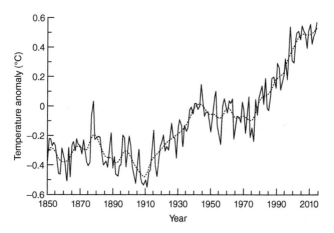

20. **Global changes in Earth's surface temperature, 1850 to 2000, expressed as differences from 1961–90 average ('temperature anomalies').**

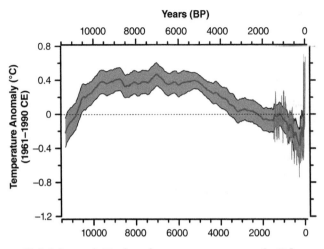

21. Global changes in Earth surface temperatures across the Holocene (temperature anomaly relative to 1961–90 mean).

have consistently demonstrated that contemporary increases in global temperatures cannot be explained by any Earth system process other than increasing anthropogenic greenhouse gas emissions. Moreover, global mean temperatures are now substantially higher than they were 100 years ago, and probably higher than they have been throughout the entire Holocene Epoch (Figure 21). And both anthropogenic greenhouse gas emissions and temperatures continue to rise together at accelerating rates. Even as you read this, it is likely that Earth is now hotter, on average, than at any other time in more than 100,000 years.

Tipping points

Major shifts in Earth's climate are the norm, not the exception, in the Quaternary, which includes dozens of glacial to interglacial transitions. Earth was also significantly warmer during the Eemian, the last interglacial interval before the Holocene, which ended about 115,000 years ago. The relatively stable and moderate interglacial temperatures of the Holocene therefore stand out as an island of climate stability within a sea of extremes. If Earth's climate system were to leave this relatively stable state, there is every reason to believe that the consequences might be catastrophic both to human societies and to non-human life as we know it. No industrial or even agricultural society has ever experienced climate shifts like those common before the Holocene. And greenhouse gas emissions and climate change are far from the only Earth system alterations that have accelerated since the 1950s.

The potential for rapid and transformative 'regime shifts' in Earth's climate is well supported by past patterns of Earth system behaviour. Glacial to interglacial transitions are one example, when warming temperatures induced by increases in incoming solar energy, reinforced by biospheric carbon emissions, decreases in sea ice and continental ice-sheet cover, and other internal positive feedbacks drive rapid changes in climate. In these cases, Earth's warming past a threshold temperature, or tipping point, triggers a self-reinforcing process of system change, resulting in a relatively rapid, non-linear, and potentially irreversible 'step-change', or regime shift in Earth's climate system. While Earth's cycling from glacial to interglacial and back again represents a 'bi-stable' system with two states, glacial and interglacial, there are also examples of one-way regime shifts in the Earth system. These include Earth's rapid cooling when the sun is blocked by dust emissions from major volcanic eruptions and meteorite impacts, and those caused by evolutionary changes in the biosphere, the most prominent having been the oxygenation of Earth's atmosphere and the emergence of life on land.

The potential for such an Anthropocene regime shift has concerned Earth system scientists for decades. The principles behind such a shift have been illustrated by the analogue of a

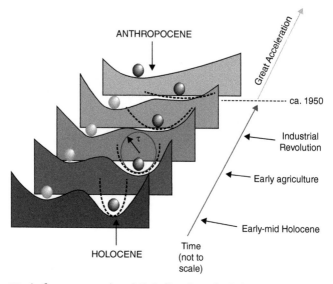

22. **Anthropocene regime shift; ball-and-cup depiction. Cup on right represents a stable basin of attraction (the Holocene) and ball on right, the state of the Earth System. Cup and ball on left represent a potential state (the Anthropocene) of the Earth System. Under gradual anthropogenic forcing, the cup becomes shallower and finally disappears (a threshold, _c._1950), causing ball to roll to the left (the regime shift) into the trajectory of the Anthropocene toward a potential future basin of attraction.**

ball rolling around within a cup. The state of the Earth system, represented by the ball, rolls around within its 'natural range' of variability, represented by the cup, or 'basin of attraction' (Figure 22). When the system is stable (early–mid Holocene), the cup is deep and narrow and the ball's movements are rapid but restricted to a narrow range. System change begins as the cup becomes more shallow and the ball is freed to shift around more broadly but also more slowly. Earlier changes in the Earth system caused by the rise of agriculture and industrial processes might have caused such increased variability. Ultimately, the system leaves, permanently, its earlier, more stable state (the Holocene), and tips into a new, less stable state (the Anthropocene). At this point, it is clear only that the Earth system is on a rapid trajectory out of the Holocene. It is still too early to know what an Anthropocene state might ultimately be like, including its relative stability and how long it might last compared with the Holocene.

As the scale and intensity of human alterations to Earth's atmosphere, hydrosphere, and biosphere increase, this undoubtedly also increases the risk of an anthropogenic regime shift to an Anthropocene state of the Earth system. For example, rapid increases in greenhouse gas concentrations might shift the planet into an extremely warm 'greenhouse Earth' state. This is especially likely if positive feedbacks kick in. For example, warming Arctic wetlands might release a burst of methane to the atmosphere, or sea ice might collapse, increasing heat-trapping by the ocean and reducing the reflection of solar energy away from Earth. There are many possibilities, including the unknown responses of a human-altered biosphere. Moreover, such a state shift might take place gradually over millennia or might be surprisingly rapid. In either case, it is highly unlikely that Earth's climate system would or could eventually shift back to its Holocene state if human pressures were lifted.

From the perspective of Earth system science, evidence that humans have forced the Earth system into a state outside its natural range of variability is fundamental to defining the Anthropocene as a new interval of geologic time. The potential for such an Earth system regime shift from Holocene to Anthropocene has support from Earth history, and also from observations on and models of Earth's functioning as a system. Nevertheless, even while the Earth system has clearly shifted outside the known conditions of the Holocene, it is still changing so rapidly that the full characteristics of a future Anthropocene state remain unknown, other than that it will certainly be warmer and sea level will be higher.

Something new

The environmental and social changes of the past half century tell a powerful story of Earth's transformation by human societies. Earth's terrestrial surface has been altered by land clearing for agriculture and settlements. Rivers have been dammed and hydrological flows re-plumbed across the hydrosphere. Humans have reshaped the biosphere by transporting flora and fauna around the world and by driving species extinct through habitat loss and overexploitation. The global biogeochemical cycles of carbon, nitrogen, and other key elements have been transformed by fossil fuel combustion, the industrial synthesis of nitrogen fertilizers, and other human activities, with impacts ranging from widespread pollution to climate change. As a result, humans have left their mark on nearly every sphere of the Earth system. Earth's climate may already have tipped irreversibly into an unprecedented state with unknown and perhaps catastrophic consequences for human societies.

With *Planet Under Pressure* and subsequent work, Will Steffen and the IGBP community established The Great Acceleration as the mainstream scientific narrative of human-induced global environmental change and linked it closely with Earth's transition to an 'Anthropocene Era'. Yet the same striking speedup that surprised Earth system scientists was already common knowledge among environmental historians—especially John McNeill. His prescient 2000 book *Something New Under the Sun* documented an unprecedented shift in the scale and intensity of social and environmental change in the 20th century together with its acceleration after 1950. Steffen, Crutzen, and McNeill later combined forces, further establishing the Great Acceleration as the leading narrative explaining the rise of humans as a 'great force of nature' and an Earth system transition to the Anthropocene after 1950.

The Great Acceleration explains the Anthropocene transition through a complex, multi-causal, narrative that weaves together human social, political, and economic changes with their diverse environmental consequences from local to global scales, including interactions among these changes across scales. While acknowledging that human alterations began long ago, the Great Acceleration asserts that human alterations of environments prior to the 20th century, though significant in some regions, remained 'well within the bounds of the natural variability of the environment' at global scales. Preindustrial societies never produced the scales or intensities of anthropogenic environmental changes needed to 'rival the great forces of Nature'. The Anthropocene began not with the rise of agriculture or even the Industrial Revolution, but only with the rise of large-scale industrial societies after 1945 and their unprecedented capacities to alter Earth's environments globally at an accelerating pace. By the middle of the 20th century, human pressures began to reach levels capable of producing an anthropogenic regime shift in the functioning of the Earth system.

In a 2016 paper in *Science*, the Anthropocene Working Group endorsed the Great Acceleration as the main scientific narrative explaining Earth's transition to the Anthropocene. With this established, the AWG turned its focus to seeking stratigraphic signatures of the key anthropogenic changes associated with a transition to the Anthropocene in the middle of the 20th century. Among the leading candidates were deposits of radioactive fallout from tests of nuclear weapons (plutonium and carbon-14), which began in 1945 and peaked around 1963 to 1964. Another popular indicator was deposits of

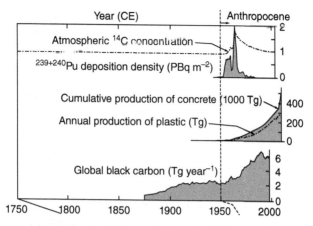

23. Novel markers of anthropogenic change including concrete, plastics, global black carbon, and plutonium (Pu) fallout, together with atmospheric radiocarbon (^{14}C) concentration.

plastics, and yet another was black carbon, produced by the incomplete combustion of fossil fuels (Figure 23). The search continues for the best stratigraphic signature and GSSP associated with the accelerating anthropogenic changes of the mid-20th century.

Chapter 5
Anthropos

'When exactly did humans attain dominance of the earth's environments?' asked archaeologists Bruce Smith and Melinda Zeder in 2013 in the journal *Anthropocene*. More than a decade after Crutzen's call to arms, archaeologists made their first bid to define the Anthropocene.

Archaeologists are the stratigraphers of the human world, specializing in reading the material records left by human societies over the long term, often from their very first beginnings. And like the stratigraphers of geology, archaeologists serve as the timekeepers of humanity, dedicated to reconstructing the social and environmental history of human societies from the physical records left behind. Over decades, their research has amassed an impressive body of evidence demonstrating that humans have dramatically altered terrestrial environments around the world starting in the late Pleistocene.

With such robust evidence that humans have transformed Earth, it might seem remarkable that archaeologists took so long to address the Anthropocene. Indeed, the call to recognize a new human epoch might have come from within archaeology itself. Yet Smith and Zeder offer good reasons why this did not happen. From their perspective as archaeologists, the onset of the Anthropocene ought not to be defined solely by the environmental consequences of human activities, but rather by the emergence of unprecedented human capacities to alter Earth's environments.

The ultimate ecosystem engineers

All organisms alter their environments merely by taking up space—even more so in feeding and sustaining themselves. But some species, known as 'ecosystem engineers', have even greater effects. These species, like dam-building beavers and burrowing earthworms, engage directly in environment-altering behaviours that profoundly change environmental conditions for themselves and others living in their environments. When these environmental alterations significantly enhance or diminish their ability to survive and to reproduce, these alterations can be considered an 'ecological inheritance'—part of an evolutionary process termed 'niche construction' by which organisms reproduce the very environmental conditions they must live within.

As Bruce Smith pointed out in *Science* in 2007, humans are the ultimate ecosystem engineers. No other single species has gained the capacity to engage in such a diverse array of potent environment altering behaviours, from land clearing using fire, to domesticating other species, to tilling the soil. This exceptional capacity to construct their own niche has helped human populations to thrive and grow beyond the natural environmental constraints that have limited other species of organisms. In the view of Smith, Zeder, and a growing number of archaeologists, an increasing human capacity for niche construction is the ultimate cause of Earth's transition to the Anthropocene.

Ancestors

The earliest records of exceptional human capacities to alter environments were laid down long before humans even existed as a species. Indeed, when *Homo sapiens* emerged about 300,000 years ago in Africa, there was little that set them apart from other species in the genus *Homo*, other than their anatomy. These first 'anatomically modern humans' were significantly less robust than their ancestors, with a lighter frame, smaller jaws and teeth, and a rounder skull. And while their brains were larger than those of prior species, they were generally smaller than those of a more robust species of *Homo* living at the same time: the Neanderthals (*Homo neanderthalensis*).

For tens of thousands of years, humans made stone tools, used fire, and lived in much the same way as their ancestors and the Neanderthals, their cousins, did. The very earliest stone tools were manufactured by our distant ancestors in the genus *Australopithecus* more than 3.3 million years ago or even before. The first stone tools manufactured by *Homo sapiens* closely resembled hand axes produced 1.6 million years earlier by our ancestors in the genus *Homo*. Controlled use of fire is well documented more than 400,000 years ago, and may have originated 2 million years ago or even before. Clearly, the 'ultimate ecosystem engineers' inherited a trick or two from their ancestors.

Gradually, starting more than 100,000 years ago, humans began making tools differently from their ancestors, using novel materials such as bones, new methods of manufacture, and more complex designs. They engraved symbolic markings into shells and bones, made and wore jewellery, and painted their bodies and cave dwellings with ochre (an iron-rich mineral) and charcoal. They traded materials needed to make tools and ornaments over long distances, including flint, obsidian, and seashells. Their settlements grew larger and more complex. And their social strategies for hunting and foraging demonstrated a whole new level of effectiveness. Over a period of a few tens of thousands of years, humans increasingly accumulated a diverse array of complex, socially learned and socially enacted 'modern' behaviours and left clear material records of these behaviours in deposits across Africa. By the late Pleistocene, more than 60,000 years ago, these complex material records began to offer evidence that new forms of 'behaviourally modern' human societies were developing social capacities beyond those of any prior species in Earth history.

A first Great Acceleration

The development and accumulation of modern human behaviours marked a major long-term shift in human capacities for niche construction. New ways of making tools, new strategies for altering and using environments, and new ways of cooperating emerged, were learned from others, and were passed on to future generations, in part through more and more sophisticated use of languages. Humans began to live in an increasingly social world where everyday survival depended on socially learned behaviours enacted in cooperation with others.

Knowledge gained by social learning—culture—became essential to harvest the right species (not the poisonous ones), to manufacture the best tools—including stone-tipped projectiles for hunting—and to alter environments in increasingly transformative ways, including the construction of traps, weirs, and rock diversions to assist in hunting and fishing. Complex social interactions and exchanges became fundamental to gaining the necessities of life, from cooperative social strategies for hunting and foraging and distributing the proceeds, to the long-distance trade of ochre, materials for stone tools (flint, obsidian), and jewellery (shells, feathers), to new strategies for exchanging food beyond the classic biological relations of kinship. As with ecosystem engineering and social exchange, ever more diverse forms of social interaction began to evolve ever more rapidly through processes of cultural evolution, including complex hierarchical relations and specialized roles, from the ceremonial duties of the shaman to the rise of larger social groupings, including tribal societies with distinct levels of leadership beyond the small egalitarian groups, or bands, that characterized the earliest human societies. The development of human languages probably played a critical role in this by increasing the fidelity by which cultural information could be transmitted socially and across generations.

As socially transmitted information, or culture, accumulated, it also evolved. By the end of the Pleistocene, human niche construction engaged a rich array of tools and techniques for living in, using, and altering a wide range of environments. Human societies also became increasingly capable of cooperative social efforts, including greater scales of coordinated activity in using and altering environments. Armed with these new social capacities, humans began their spread out of Africa in multiple waves of emigration more than 60,000 years ago, carrying with them the social capacity to change the world (Figure 24). While the first behaviourally modern human societies probably developed within Africa, their spread across Earth would soon make *Homo sapiens* a global species. Fourteen thousand years ago, before the Pleistocene had ended and the Holocene began, human populations were established on every continent except Antarctica.

Defaunation

Human societies left archaeological evidence of their arrival wherever they went, with deposits of charcoal, tools and other artefacts, and the bones and other remains of themselves and the species they hunted and foraged for. New tool designs and ways of living emerged as different societies in different regions adapted to new environments. Hunter-gatherer societies learned to capture and consume an ever wider diversity of new species; broadening the human niche. These new species included many that had never evolved to live with a tool-wielding, fire-using, social-hunting, niche-constructing primate. Many of these species, especially the larger animals prized as human prey—the megafauna—would soon be extinct, from the Glyptodon, a giant relative of the armadillo, to the elephant-sized ground sloths of the genus *Megatherium*.

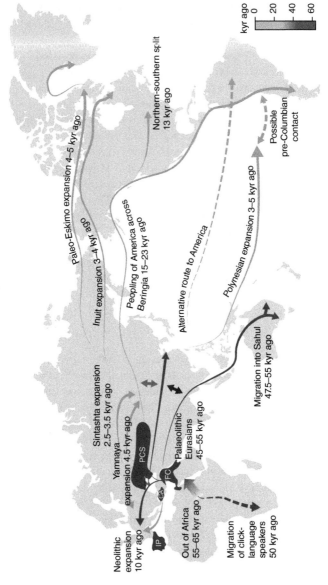

24. Map of the human diaspora out of Africa illustrating multiple migrations over time across the continents.

In the Late Pleistocene and Early Holocene, hunter-gatherers probably caused the extinction of about half of Earth's large-bodied mammals and a number of large bird species in Australia. The Americas and Australia lost the most species—from 70 to almost 90 per cent of all their mammalian megafauna—while Eurasia lost less than 40 per cent and Africa only about 20 per cent. It is likely that these regional differences reflect their history of prior exposure to the genus *Homo*; fauna in Africa and Eurasia co-evolved together with humans. In the Americas and Australia, megafauna were not so lucky; with no prior experience they suddenly faced Earth's most successful predator, armed with projectile weapons, fire, and the ability to coordinate hunting strategies across large groups (Figure 25).

The degree to which human hunter-gatherers caused the mass extinction of megafauna remains a subject of ongoing debate among scientists. One of the most intriguing megafauna

25. **Social hunting of woolly mammoth using stone-tipped projectiles.**

extinctions is that of our closest relatives, the Neanderthals, who coexisted with and even bred with us for thousands of years following human arrival in Eurasia, becoming functionally extinct only about 40,000 years ago. Competition is one explanation, another is disease, and another is climate change. Climate changes are held at least partly responsible for many of the megafauna extinctions, as cold intervals in the late Pleistocene and early Holocene co-occurred with human arrivals. Yet these same species had already outlasted dozens of glacial/interglacial cycles without extinction. Probably, climate change combined with human predation was an especially potent driver of extinction. Human use of fire may also have contributed, as this may have increased the frequency and extent of burning across drier regions, altering and shifting natural habitats as an unintended consequence of fire use.

Contemporary hunter-gatherers also use fire intentionally to open up dense forests and increase the productivity of ground vegetation, which attracts game and increases success in hunting and foraging. Early hunter-gatherers might have used similar practices to reshape vegetation across landscapes. Moreover, the removal of mega-herbivores and mega-carnivores also reshapes vegetation growth. For example, human-driven extinction of the woolly mammoth may have reduced pressures on woody vegetation, leading to woody regrowth across the vast northern grasslands of the 'Mammoth Steppe'; both grasslands and mammoths disappeared at approximately the same time. There is even plausible evidence that loss of the seed-dispersal capabilities provided by megafauna—some trees produce seeds far too large to be dispersed by smaller species—may have limited regeneration of many highly productive tree species, reducing carbon uptake by forests.

A megafauna Anthropocene?

A robust body of evidence now confirms that human hunter-gatherers dramatically altered patterns of flora and fauna across Earth in the last 50,000 years of the Late Pleistocene and at the beginning of the Holocene. There is also evidence suggesting that loss of megafauna and use of fire might have altered vegetation cover across the continents to such a degree that this significantly altered Earth's climate. Rapid increases in woody vegetation take up atmospheric carbon dioxide, causing cooling. Yet dense woody vegetation, like the ocean, is also darker than barren and snow-covered areas, and therefore causes warming by taking up more of the sun's energy. This warming is enhanced further by vegetation exchanges of moisture and energy with the atmosphere. There is even limited evidence that methane release from the digestive systems of megafauna might have significantly warmed Earth—such that their loss might have cooled the planet near the end of the Pleistocene.

Multiple scientific proposals now aim to recognize the global environmental consequences of human-driven megafauna extinctions and enhanced fire regimes, especially across the Americas, as the basis for an Anthropocene start date near the end of the Pleistocene, approximately 14,000 years ago. These long Anthropocene proposals, while both intriguing and suggestive, are challenged by multiple deficiencies. There is little doubt that human-driven megafauna extinctions altered ecological functioning across multiple continents with significant repercussions for the functioning of the terrestrial biosphere. Nevertheless, evidence supporting the claim that these also altered Earth's climate and functioning as a system, both empirically and using climate simulation models, remains far from levels needed to convince most scientists. In addition, human-caused megafauna extinctions and shifts in vegetation resemble those occurring prior to humans and were highly diachronous. It is therefore highly unlikely that these changes will ultimately enable the identification of a golden spike that can be correlated in time across different sites around the world.

Agriculture

As human hunter-gatherers expanded their niche construction across Earth, they began to transform the biosphere. Yet these early transformations pale before those that came next. According to Smith and Zeder, it is the 'domestication process...that provides the archaeological signature for major human manipulation of terrestrial ecosystems, and the onset of the Anthropocene'. The rise and spread of agricultural societies unleashed a global process of transformative environmental change that continues to this day.

Agriculture evolved from the pre- and proto-agricultural niche construction practices of hunter-gatherers. As populations grew and culture accumulated, hunter-gatherers developed a diverse array of socially learned behaviours that enabled them to enhance the productivity of their environments and also helped them to adapt to the environmental changes introduced by their ancestors. Once preferred megafauna and other species became rare or extinct, hunter-gatherers learned to harvest more species, broadening their diets and their niche. They burned vegetation to encourage new growth. They learned to increase the nutritional returns from hunting and foraging by cooking, grinding, and efficiently processing animal and plant foods, making small grains and tubers useful for the first time. They spread the seeds of the plant foods they liked and managed the populations of the animals they hunted—and would later domesticate. These niche construction practices were much less productive than the agricultural technologies that came later, but they still enabled human populations to grow well beyond those supportable on unaltered ecosystems. Increasingly intensive environmental alterations would be required to support these growing and increasingly complex societies, together with out-migrations to less populous regions. The stage was set to take human niche construction to a whole new level; the rise and spread of agriculture.

Societies dependent on agriculture emerged in more than a dozen centres of domestication on every populated continent except Australia (Figure 26). Some developed at the Pleistocene–Holocene transition, as in Southwest Asia, South America, and North China, others closer to 6,000–8,000 years ago, like Yangtze China and Central America, while still others developed 4,000–5,000 years ago in Africa, India, Southeast Asia, and the North American prairies. In some cases, sedentary hunter-gatherers transitioned to farming, as in Southwest Asia or Yangtze China; in others, mobile hunters took up herding, as in Africa, or mobile hunter-gatherers engaged in mobile forms of agriculture like shifting cultivation, as in India, New Guinea, and South America (Figure 27).

Agricultural populations grew more rapidly than those of hunter-gatherers and ultimately displaced them across Earth's most productive lands both directly and when hunter-gatherers adopted agricultural practices themselves. The social and environmental changes brought by agriculture were far from linear over time. Countless societies collapsed and began anew. Still, there is a clear long-term trend towards ever-larger scales of agrarian societies supported by increasingly productive land use practices, or 'land use intensification', over time. Early practices of shifting cultivation used land for a year or two and then cleared more once soil fertility declined. Agrarian populations, resource demands, and social and cultural capacities grew and developed, and more labour- and energy-intensive techniques were adopted to increase the productivity of land, including the planting of crops every year, irrigation, manuring, the plough, and other methods. Agricultural intensification using manures began in Southwest Asia and Europe as early as 8,000 years ago, based on stable nitrogen isotope ratios in preserved grains, and irrigated rice

Anthropocene

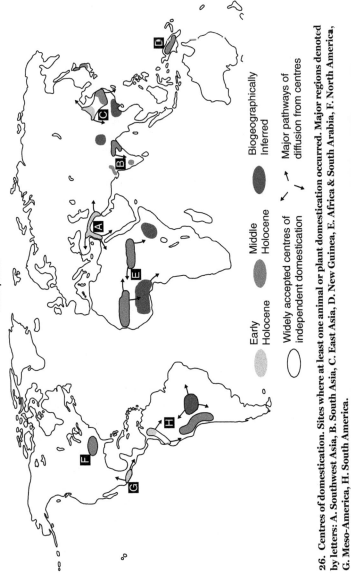

26. Centres of domestication. Sites where at least one animal or plant domestication occurred. Major regions denoted by letters: A. Southwest Asia, B. South Asia, C. East Asia, D. New Guinea, E. Africa & South Arabia, F. North America, G. Meso-America, H. South America.

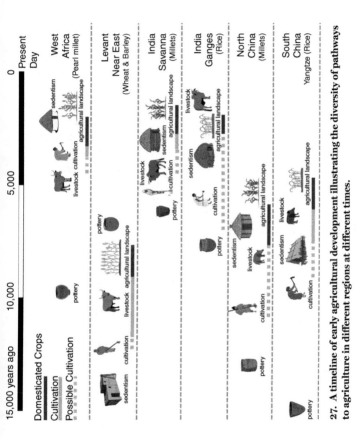

27. A timeline of early agricultural development illustrating the diversity of pathways to agriculture in different regions at different times.

production is in evidence as early as 7,000 years ago in China and India, becoming widely established in key rice growing regions by about 5,000 years ago.

Abundant evidence confirms that agricultural land use was widespread by the mid-Holocene, including deposits of soils and sediments caused by increased soil erosion, charcoal, the remains of crop plants and weeds, including pollen, starch grains, and phytoliths (silica crystals produced in plant cells), the bones and other remains of domestic livestock, changes in the isotopic composition of soils and fossil manures, and long-term changes in vegetation structure and species composition left behind after early land clearing and soil tillage; present-day woodlands from the Mediterranean to the Tropics are increasingly recognized as the bio-cultural legacies of long histories of prior human use. Agricultural land use also produced anthropogenic soils, from the manure-enriched 'plaggen' soils of north-western Europe, which may date to 4000 BC, to the 'terra preta', or 'dark earth' soils enriched with charcoal and waste materials that are observed across the Amazon basin dating perhaps to 500 BC and may also have been produced in Africa, together with various 'anthrosols' altered by manuring, tillage, irrigation, and other land use practices in different regions. The widespread presence of anthropogenic soils has been suggested as a golden spike for the Anthropocene, *circa* 2,000 years ago, though the prospects for a successful GSSP proposal based on anthropogenic soils are not strong, owing to their diachronous origins.

Agricultural transformation of Earth began more than 10,000 years ago and continues to convert natural habitats to agricultural landscapes engineered and managed to support populations of domesticated species (Figure 28). Spreading gradually across the continents over millennia, agriculture began leaving a legacy of altered soil chemistry and sedimentary processes transformed by land clearing, soil tillage, and erosion. Hydrology was altered by reservoirs and irrigation systems. And the functioning of the

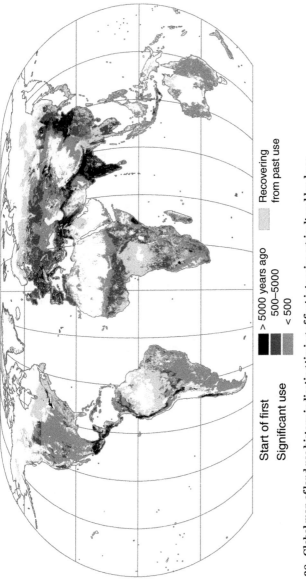

Start of first

Significant use

■ > 5000 years ago
■ 500–5000
■ < 500

□ Recovering
from past use

28. Global map of land use history indicating timing of first intensive agricultural land use.

biosphere, the atmosphere, and the Earth system as a whole began to change.

The Early Anthropogenic Hypothesis

In 2003, climate scientist William Ruddiman published a paper entitled 'The anthropogenic greenhouse era began thousands of years ago'. Ruddiman claimed that by clearing forests for agriculture and irrigating rice paddies, ancient farmers produced emissions of carbon dioxide and methane sufficient to significantly alter greenhouse gas concentrations in Earth's atmosphere (Figure 29). In so doing, they caused a warming effect adequate to delay the next glacial cycle of the Quaternary. Ruddiman's 'Early Anthropogenic Hypothesis' has since been put to the test in dozens of research papers. Some of its claims remain controversial. Yet its basic premise, that human societies gained the capacity to transform Earth's functioning as a system long before the industrial era, remains under serious consideration by Earth system scientists and is supported by multiple lines of evidence.

Ruddiman's hypothesis compares the 'natural' downward trends in atmospheric carbon dioxide and methane observed in prior interglacial intervals with those of the Holocene. Unlike prior interglacials, methane concentrations stopped declining in the mid-Holocene, 5,000 years ago, and began to rise. A similar trend is observable in carbon dioxide as well, starting 7,000 years ago. Ruddiman's hypothesis ascribes these anomalous trends to greenhouse gas emissions caused by agricultural use of land.

In 2011, archaeologist Dorian Fuller used a historical model of rice areas to demonstrate that methane emissions from early rice production could account for about 80 per cent of the early anthropogenic trend in atmospheric methane (Figure 30). Subsequent work using carbon isotopes has confirmed that these early methane emissions were indeed anthropogenic.

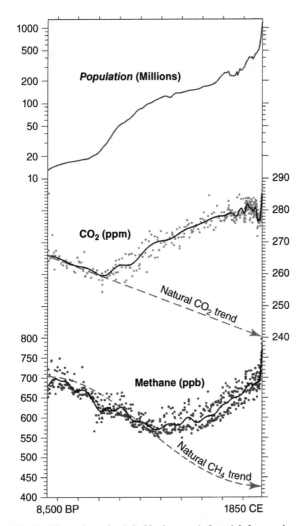

29. The Ruddiman hypothesis holds that pre-industrial changes in atmospheric CO_2 and methane in mid-Holocene deviated from 'natural' trends observed in prior interglacial intervals and that these deviations are caused by agricultural land clearing (CO_2), and by rice farming and domestic ungulates (mostly cattle and buffalo) in the case of methane.

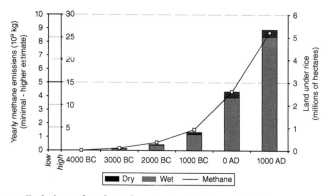

30. Emissions of methane from land under dry and wet rice production from 6,000 to 1,000 years before present (4000 BC to AD 1000).

Accounting for the anomalous trend in carbon dioxide has been more challenging, in part because the global biogeochemical fluxes that regulate carbon dioxide in the atmosphere are more complex and harder to measure than those for methane. For example, it is necessary to consider, simultaneously, the rates of carbon emission when land is cleared, vegetation is burned, and soil is tilled, as well as carbon uptake when vegetation regrows after land is abandoned. It is also necessary to balance emissions with carbon uptake by the ocean, peatlands, and other parts of the global carbon cycle.

Critics have asked how the tiny agricultural populations of the mid-Holocene, especially the few tens of millions present 7,000 years ago, could have cleared and farmed such a relatively large area of land, and why their emissions do not appear to accelerate with population growth later in the Holocene. If emissions are estimated based on a constant amount of land used per person, early land clearing and carbon emissions are much too small to support Ruddiman. But when historical trends in land use intensification are included, with early farmers using much more land per person than farmers do today (though far less intensively), land clearing and emissions follow trends similar to those observed in the mid- to late Holocene.

Recent climate simulations have confirmed that greenhouse gas emissions by early agricultural societies had the potential to alter Earth's climate trajectory, though the amount of this change and its timing remains a subject of active research. Major agricultural emissions of carbon dioxide from about 7,000 years ago remain a plausible if controversial hypothesis. However, early methane emissions from rice production are now widely accepted as the cause of a substantial rise in atmospheric methane concentrations approximately 5,000 years before present. Marked in an ice core sequence similar to the one defining the Holocene, early anthropogenic changes in atmospheric methane have been proposed as having the potential to serve as a golden spike marking the lower boundary of the Anthropocene, though difficulties of correlation globally beyond the ice core pose difficulties for its adoption as a formal chronostratigraphic boundary in the GTS.

Scaling up and out

Agricultural populations continued to grow, increased in density, and had spread across all continents except Australia by 6,000 years ago, despite the periodic collapse of individual societies. Increasingly intensive and productive land use systems evolved in support of ever denser populations. The higher yields of these systems produced agricultural surpluses that were extracted by trade and taxation, enabling the rise of urban populations with increasingly hierarchical and complex societies with specialized roles from artisan to trader to king, and new tools for living in a social world, including money, writing, and metallurgy—the key to new forms of weaponry. Economies of scale in larger urban populations offered multiple benefits, including increased access to wealth and services, and attracted rural people, driving further urban growth—with the exception of periodic outbreaks of disease. The first cities with populations of 50,000 or more became the power and trade centres of larger-scale societies, beginning by about 3000 BC in the Indus Valley. By AD 1, cities with populations of hundreds of thousands thrived across the Near East, Europe, and Asia, and became increasingly dependent on extensive networks of trade, some of which stretched across continents, like the silk road connecting Western Europe with eastern China, and also by a growing number of maritime routes. Human societies were scaling both up and out.

Through trade, warfare, religion, and other social interactions, human societies became increasingly interconnected into 'world systems' of exchange. Cultural knowledge, artefacts, natural resources, and living organisms spread rapidly across these world systems, both intentionally as trade goods, and unintentionally as stowaways, including pests and diseases. Roads and waterways were constructed to transport goods over long distances. Societies explored, expanded to, and traded with new lands and new societies, supported by increasingly seaworthy watercraft and navigational techniques enabling the open seas to become highways of inter-societal exchange.

Even with more traditional technologies, Polynesian societies were able to spread across the islands of the Pacific by boat starting about 3,500 years ago, carrying with them a suite of domesticated species, from bananas and yams to dogs, pigs, and chickens, and also, inadvertently, rats. The colonization of new lands by a complex agricultural society profoundly transformed landscapes and ecosystems by fire, land clearing, the cultivation of domesticates, and the introduction of rats and other new species of animals and plants which both consumed and outcompeted large numbers of native species. Megafauna, smaller animals, and even many plant species went extinct. The classic material evidence of agricultural colonization, in the form of cultural artefacts, charcoal, eroded soils, new species, and widespread species extinctions, is evident across the Pacific from Hawaii to New Zealand.

A global world system

Though the 'Old World' societies of Eurasia were already interconnected by exchange more than 2,000 years ago, human societies were not yet globally connected, despite their presence on every continent except Antarctica. Increasing European demands for new wealth, power, and influence would change this, by driving an expansion of trading efforts beyond customary routes. Ultimately, more than 500 years ago, these efforts would result in the first substantial two-way exchange of culture and biology between Europe and the Americas. The accidental 'discovery' of the Americas by Christopher Columbus set off a process of global social and environmental change like no other before, the Columbian Exchange, through which the Old World and the New World became one. Driven by European efforts to extract wealth from the Americas, human societies were integrated for the first time into a truly global world system of social, material, and biological exchange.

Europeans coveted gold, spices, and other rare natural resources, but their trade routes also carried a host of profoundly transformative social and biological forces: new cultural practices, technologies, domesticates, and diseases. American potatoes, tomatoes, chilli peppers, and maize transformed farming systems around the world, not just in Europe, but in Asia and Africa too. Domestic livestock, from horses (native horses were lost in the Pleistocene extinction), to cattle, to pigs, changed livelihood strategies across the Americas. Many other species came along for the ride, mixing together populations of flora and fauna that had evolved separately on different continents for millions of years, in a rapid process of 'biotic homogenization'. All of these changes left evidence in the stratigraphic record, but one specific biological exchange stands out for its rapid transformative effects.

The introduction of smallpox and other Old World diseases is estimated to have killed 50 million native Americans between 1492 and 1650 in epidemics of European diseases to which they had never before been exposed. The results were catastrophic, with whole societies collapsing in the face of rapid population declines of 50 to 90 per cent and more. Epidemics spread so rapidly through indigenous exchange networks that many native societies were wiped out before Europeans first reached them. Forced labour, resettlement, colonial violence, and imported slaves only accelerated the decimation. Before Europeans began transforming American landscapes into large-scale commercial plantations and ranches, the indigenous societies that had long cultivated crops and used fire to manage their vegetation had shrunk to a tiny proportion of their former extent. In their absence, forests began to regrow, taking up so much carbon in the process that they could have significantly reduced atmospheric carbon dioxide, an effect that may be evident in ice core measurements around 1610.

In 2015, an ecologist and a geographer, Simon Lewis and Mark Maslin, in a *Nature* review of Anthropocene GSSP proposals, also introduced their own: 'the Orbis spike' (Figure 31). Orbis, Latin for world, proposed that the Anthropocene was initiated by the Columbian Exchange, in which the 'collision of the Old and New Worlds' marked humans as not only a global species, but now also a global system and a global force with geologically unprecedented consequences, including the global interchange and homogenization of Earth's biota. Moreover, the unprecedented scales of social change, resource extraction, and commercial land use unleashed by Europeans in the Americas ultimately fuelled the development of industrial societies. The emergence of Earth's first global human system unfolded over hundreds of years, leaving a permanent, though mostly diachronous, record in the global homogenization of flora and fauna together with the usual material evidence of transformative social-environmental change. There was, however, at least one rapid global change

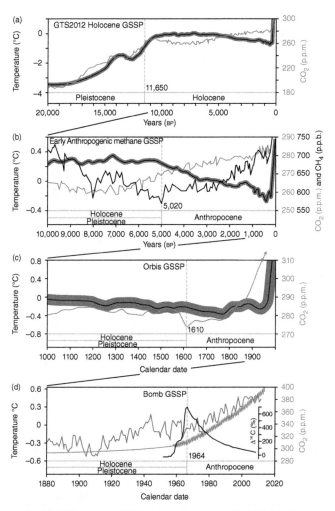

31. Early Anthropocene GSSP proposals compared with (a) Holocene, based on CO$_2$ changes relating to glacial/interglacial transition 11,650 years before present, (b) Anthropogenic methane increase 5,020 years before present (Ruddiman hypothesis), (c) Orbis 'spike' in CO$_2$ *circa* 1610, and (d) Peak radiocarbon (^{14}C) levels in tree rings *circa* 1964 induced by atmospheric nuclear bomb tests.

that could potentially define an Orbis GSSP for the Anthropocene: a small dip in carbon dioxide concentrations around 1610, marked in an ice core.

Human time

As archaeologist Matthew Edgeworth and others pointed out in a paper entitled 'Diachronous Beginnings of the Anthropocene', 'archaeology and geology are related disciplines' and rely on largely the same stratigraphic principles. Geologists and archaeologists often work together in the same places, with geologists focusing on the natural processes that have shaped a site over time, and archaeologists determining the lower boundaries where humans have established their own layers of anthropogenic material deposits—what Edgeworth has termed the 'archaeosphere'. In other words, the archaeosphere might be considered a dividing line between the stratigraphic expertise of geologists and archaeologists.

The nature of the material record studied by archaeological stratigraphers is especially complex, heterogeneous, and diachronous (Figure 32). It is common for the deposits made by one society, or even just one household, to be reworked by another through the digging of ditches, foundations, and graves, to which are added still further buildings, rubbish, and debris, all of which might later be covered by layers of sediment deposited through flooding and other natural processes, or have been cleared away to begin construction anew. It may show correlations in its depth and composition among the sites produced by a given society, or it may not. It may be pierced in one place by catacombs, deep wells, and subway tunnels, in another, covered by tilled soils, artificial wetlands, a landfill, or a hill composed of multiple layers of settlements laid down one over another over millennia (a 'tell', a common archaeological feature in the Middle East). It ranges from non-existent in some places to tens of metres deep in others. At every scale, from site, to region, and especially globally,

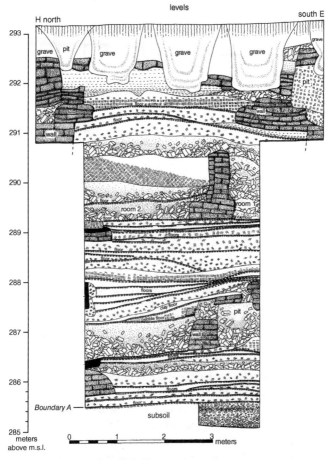

levels

H north   south E

293 —

292 —

291 —

290 —

289 —

288 —

287 —

286 —

285 —
meters
above m.s.l.

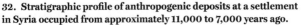

0   1   2   3
meters

32. Stratigraphic profile of anthropogenic deposits at a settlement in Syria occupied from approximately 11,000 to 7,000 years ago.

the archaeosphere is intensely heterogeneous and diachronous. In Edgeworth's view, and that of archaeologists in general, diachroneity defines not only the archaeosphere, but also the Anthropocene itself.

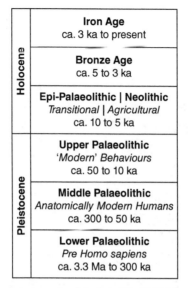

	Iron Age ca. 3 ka to present
Holocene	Bronze Age ca. 5 to 3 ka
	Epi-Palaeolithic | Neolithic *Transitional | Agricultural* ca. 10 to 5 ka
Pleistocene	Upper Palaeolithic *'Modern' Behaviours* ca. 50 to 10 ka
	Middle Palaeolithic *Anatomically Modern Humans* ca. 300 to 50 ka
	Lower Palaeolithic *Pre Homo sapiens* ca. 3.3 Ma to 300 ka

33. **The three age system of archaeology. The 'lithic' ages represent the 'stone age', differentiated from the bronze and iron ages. These general patterns of cultural change are not always observed and are also supplemented by more detailed time divisions relevant to specific sites and regions.**

Archaeologists also use stratigraphic methods to produce calendars of human time (Figure 33). But unlike geologic time, even the most generalized archaeological calendars are diachronous *by design*. Their goal is to characterize the different developmental pathways taken by different societies, in different places, at different times. There is a general system of archaeological 'ages', the first of which begins with the first stone tools in the Palaeolithic, or 'old' stone age, which ends with the Pleistocene. The Holocene begins with Epi-Palaeolithic societies, which largely continued Palaeolithic lifeways, and Neolithic societies, which adopted agriculture. Bronze and Iron Age societies are recognized by the capacity to produce these metals and concomitant shifts in societal scale and complexity. Yet despite some remarkable parallels in their development, Neolithic societies emerged around the world in different places and at different times, like those across the Americas, the Middle East, and East Asia. Archaeologists also depend on far more detailed local and regional time systems to interpret the developmental periods of different societies. In archaeology, there is no goal of producing a globally synchronous timeline of human social change, or its impacts on environments, because this is not the way that human societies have formed or how they change.

Thicker and deeper

The Anthropocene tells a story about human capacity to transform Earth. But when does the story begin? The AWG has focused on the middle of the 20th century as the most suitable for a GSSP marking the start of the Anthropocene Epoch in the GTS—while noting that anthropogenic influence began far earlier. But for archaeologists, anthropologists, geographers, geologists, and others focused on the long-term causes, rather than the consequences, of anthropogenic global environmental change, the Anthropocene begins long before 1950.

An earlier Anthropocene might recognize the Late Pleistocene megafauna extinctions, the emergence and spread of agriculture, rising atmospheric methane from rice production 5,000 years ago, widespread anthropogenic soils 2,000 years ago, the formation of a global world system *circa* 500 years ago (the Orbis spike), or the start of the industrial era *circa* 200 years ago. Some of these alternative proposals for earlier GSSPs include stratigraphic evidence, such as the signals in ice cores mentioned earlier, though the AWG has found them insufficient to satisfy the stratigraphic criteria used to construct the GTS.

Smith and Zeder argued that there is no need for a new GSSP. The Holocene might simply be renamed as the Holocene/Anthropocene. Alternatively, the origins of human transformation of Earth might be recognized by a non-geological time interval; a 'Palaeoanthropocene'. Others, including Ruddiman, have proposed that because of the continuous nature of anthropogenic environmental change, the Anthropocene should not be formalized at all, but used informally, as a lower case 'anthropocene'. The one thing that unites all of these proposals is their common focus on recognizing the long, rich, and diachronous history of human transformation of Earth's environments. The Industrial Revolution and the Great Acceleration are merely the latest, and most impactful, chapters in a long, entangled, and evolving history of human transformation of Earth's environments; a history that is still unfolding.

The stratigraphers of the human world have established that the processes by which human societies and their environments change and evolve are cumulative, continuous, heterogeneous, diachronous, and complex. The material evidence of anthropogenic environmental transformations is equally complex and diachronous, stretches deep into the human past, and is spread broadly across Earth. From an archaeological point of view, there is nothing recent or unusual about human alteration of Earth's environments. The human world has always been anthropogenic. Nearly every human society in Earth history has lived in environments transformed by their ancestors.

Early human transformation of Earth, while clearly much smaller in scale and less rapid than today's, has left evidence just as permanent as the evidence deposited later—it is just buried deeper and scattered more widely across the sands of time. But it is precisely this gradual build-up of anthropogenic layers and features from prehistory to the present day that archaeologists find important to study, not the identification of a precise boundary in time and rocks marking globally significant human transformation of Earth.

Chapter 6
Oikos

Human reshaping of ecology has been driving the Anthropocene transition since its very first beginnings. Mass extinction and species invasions, greenhouse gas emissions, climate change, altered soils and hydrology, the massive conversion of natural habitats to anthropogenic landscapes—all were produced by anthropogenic ecological change. The ecological and environmental sciences have been instrumental in characterizing these changes, yet they have also struggled to understand them as more than just a temporary disturbance of an otherwise natural world. For example, the Anthropocene poses even greater challenges for those working to conserve and restore natural habitats. What does 'natural habitat' even mean on a planet transformed by humans? In a controversial 2011 essay entitled 'Conservation in the Anthropocene', Peter Kareiva, then chief scientist of one of the world's largest conservation organizations, The Nature Conservancy, summed it up this way:

> the global scale of this transformation has reinforced conservation's intense nostalgia for wilderness and a past of pristine nature. But conservation's continuing focus upon preserving islands of Holocene ecosystems in the age of the Anthropocene is both anachronistic and counterproductive.

Even as the results of ecological science have helped to characterize the Anthropocene, ecology as a discipline has been reshaped by the need for new approaches to address the ecology of an Earth transformed by human societies. New paradigms have emerged, redefining the value of nature and the role of humans in shaping and curating the ecology of an increasingly anthropogenic biosphere.

Dividing nature

Ecology, from the Greek 'oikos' ('house'), is a relatively new and integrative scientific discipline focused on understanding interactions among organisms and their environments, including the 'food chains' that connect carnivores, herbivores, and plants; the spatial patterns of plant and animal populations; and the biogeochemical fluxes among organisms and their abiotic environments. Emerging at the end of the 19th century, ecology has deep roots in natural history dating back to Aristotle and before. Charles Darwin was a naturalist, as were Carolus Linnaeus (1707–78), who classified life into species, and Alexander von Humboldt (1769–1859), who mapped life's global environmental patterns.

Darwin and most other naturalists were comfortable with including humans in their work—at least prehistoric humans and their non-European contemporaries. But this had already begun to change in the late 18th century, when Comte de Buffon distinguished between 'original nature' and 'nature civilised' by humans. This division into human and non-human natures deepened with the rise of the natural sciences, ecology among them, which left the study of the human world to the social sciences and the humanities. Ecologists, like archaeologists and anthropologists, developed traditions of studying smaller sites and regions, where regional and global interactions of humans and the natural world might be considered external to their studies.

Dividing nature into two parts has always been problematic, especially given who is doing the dividing. Yet this unnatural act of division may also have heightened the sensitivity of ecologists to the transformative capacities of human societies. In 1778, Comte de Buffon was already prepared to claim that 'the entire face of the Earth bears the imprint of human power'. In 1997, ecologist Peter Vitousek and colleagues published a hugely influential *Science* paper offering evidence 'that we live on a human-dominated planet'. And the first person to name the Anthropocene was not Paul Crutzen, but lake ecologist Eugene Stoermer.

The Pristine Myth

To study habitats and ecosystems uninfluenced by humans, many ecologists, especially in North America, have sought out places without clear evidence of human activity. Yet even before anthropogenic climate change became too pervasive to ignore, this strategy was already scientifically suspect.

Oikos

Palaeoecologists, the stratigraphers of ecology, reconstruct the ecological changes of the past from the material remains of past ecosystems. Together with archaeologists, palaeontologists (fossil specialists), environmental historians, and others, their work has established that human transformation of ecosystems has produced ecological legacies that have lingered from the Late Pleistocene to this day (Figure 34). Extinctions of large herbivores like the woolly mammoth transformed grasslands into woodlands. The management of vegetation using fire disturbed soils and altered nutrient levels. Even the earliest agriculture rearranged nutrients across landscapes, increasing soil fertility in some places and decreasing it in others, permanently altering soil chemistry and other soil properties. These anthropogenic legacies in soils can still shape species composition and the productivity of plant communities centuries and even millennia later. Species have also been redistributed across landscapes and regions by trade, migrations, and the intentional efforts of hunter-gatherers,

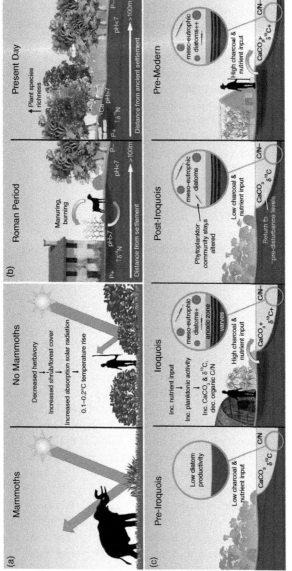

Anthropocene

34. Long-term ecological transformation of landscapes illustrating (a) impact of eliminating large herbivores, (b) long-term effects of ancient agriculture on soil geochemistry and plant biodiversity in forests, and (c) responses to cultural disturbance of lake watershed.

farmers, and traders. Long-term stratigraphic records of these human influences have accumulated in lakes, ponds, wetlands, and other low-lying areas within landscapes, both directly as deposits of charcoal, pollen, and sediments with changing chemistry and isotopic signatures, and indirectly, as diatoms (microscopic algae) and other water plants responded to changes in nutrients and other external influences.

Even in many regions where human influences appear to be absent, palaeoecological evidence regularly demonstrates that contemporary ecological patterns and processes were shaped by earlier human societies. Just imagine what the vegetation of northern Europe or Canada would look like if the woolly mammoth were still around. Even remote Amazonian rainforests have been shown to have tree distributions shaped by human efforts to disperse the most desirable species, like the Brazil nut, which has been propagated by rainforest people for thousands of years, and is still mostly harvested from wild trees. Evidence continues to grow that the vast tropical rainforests of Amazonia and the Congo may have been largely reshaped by human use of fire, shifting cultivation, dispersal, and propagation of favoured species, and other land use practices of hunter-gatherers and early farmers. Yet many ecologists and conservationists have tended to perceive habitats without people as places without human influence.

In most of Europe and Asia, and in parts of Africa, landscapes are generally too populous and human-altered to misinterpret in this way. But the descendants of European settlers in the Americas and Australasia have regularly mistaken dense woodlands for untouched 'pristine' habitats when they are in fact still recovering from long-term management by earlier societies. Geographer William Denevan highlighted this error in his aptly titled 1992 article 'The Pristine Myth: The Landscape of the Americas in 1492'. Tim Flannery did the same for Australian landscapes in his book *The Future Eaters*. Before the Holocene had even begun, huge regions on every continent were transformed by human activities. The Pristine Myth—that places without humans today represent an ecology without prior human influence—is now recognized as a serious barrier to understanding contemporary ecological patterns and processes.

Disturbance

Cores of dated sediments extracted from lakes provide some of the most robust records of long-term ecological change. One such core from Crawford Lake in Ontario, Canada, assessed with the help of Eugene Stoermer, no less, has served as a prime example of the complexities of anthropogenic ecological change in discussions among archaeologists, geologists critical of Anthropocene formalization, and the AWG (Figures 35 and 34(c)). The core records more than 1,000 years of ecological change around the lake, with nutrient inputs from agriculture increasing algal productivity and organic carbon, and changing populations of different diatom species.

Agricultural land use left deposits of pollen from maize and common field weeds, and fungal spores from corn smut—a disease of maize. The core shows clearly that Iroquois maize farming and settlement around the lake from AD 1268 to 1486 eroded soils and increased nutrient inputs to the lake. Farming then ceased until 1867 when European settlers (Canadians) colonized the area again, farming maize and polluting the lake. Stratigraphic signals of land use change are also evident in the mid-20th century.

The complexities of human disturbance are readily apparent in the dynamic palaeoecological records of Crawford Lake and many others. In some lake deposits, but certainly not all, a radiogenic fallout layer from the mid-20th century may coincide with specific biotic or chemical changes, providing a strong basis for AWG to establish a globally correlatable marker for an Anthropocene Epoch. Either way, for ecologist Eugene Stoermer and other

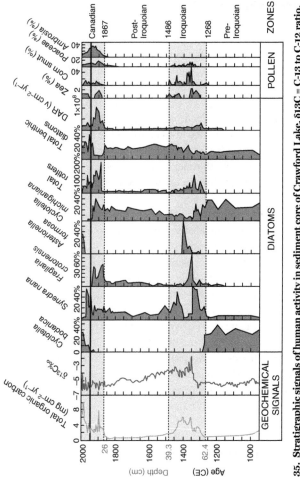

35. **Stratigraphic signals of human activity in sediment core of Crawford Lake. δ13C = C-13 to C-12 ratio.**

scientists focused on the palaeoecological records of anthropogenic change, there is no shortage of evidence that human disturbance of communities and ecosystems is complex, dynamic, diachronous, and sustained over long periods of time.

The dynamics of ecological change are complex even without humans in the mix. A prime example is fire. In drier regions, fires recur periodically, leaving deposits of charcoal, nutrients, and mosaics of vegetation patches in different stages of secondary succession. In these regions, fire forms a regular part of ecosystem functioning, defining a 'disturbance regime' of recurring fire to which many species developed adaptations like fire-retardant bark and fire-dependent seed germination. For example, cones of the Jack pine of North America open to release their seeds only when exposed to the intense heat of forest fires.

Before ecologists understood the significance of disturbance regimes, they recommended suppressing fires to maintain existing vegetation. In response, fire-adapted species failed to reproduce and flammable biomass accumulated over time, yielding fires that could not be suppressed and burned far more intensely than ever before; sometimes even soils were combusted in such fires. Ecologists learned a hard lesson: disturbance plays an important role in ecosystems and communities, and suppressing disturbances can disrupt communities and habitats. Moreover, in places like Australia and eastern North America, the fire regimes that shaped ecology across landscapes for thousands of years were anthropogenic, the product of hunter-gatherers and farmers who intentionally managed vegetation using fire.

The complex dynamics of human–environment interactions make it a challenge to detect whether a significant ecological change has occurred. To make this possible, it is necessary to characterize the 'historic range of variability' of ecological parameters, including variations in the populations of different species, abiotic environmental conditions, and frequencies of fire and other disturbances over time. By establishing this historical range as a reference or 'baseline' state, changes outside the range provide evidence of ecologically significant change.

A sixth mass extinction

Species extinctions are among the most significant ecological changes human societies have yet produced. Their causes are many. Overexploitation dates to the Pleistocene and remains important. Land use for agriculture and settlements has long been and continues to be the most potent and ongoing driver of terrestrial extinctions. By shrinking, fragmenting, and transforming habitats, land use reduces the resources available to vulnerable populations and divides them into smaller, less viable groups, increasing the probability of extinction. Introductions of non-native species have also played a major role in native extinctions, especially on islands, where endemic species—species that are restricted to local areas, sometimes just a single small island—have proved most vulnerable of all. Rats, pigs, dogs, cats, and other introductions have devastated species without evolved defences to them. The famed extinctions of the Dodo and the trees of Easter Island, once attributed solely to overexploitation, are now considered largely the result of introduced species feeding on their eggs and seeds, respectively. More recently, toxic pollutants, including the pesticide DDT, have put species, especially those at the top of the food chain, at risk of extinction. Anthropogenic global climate change is now emerging as what may become the most significant extinction driver of all time, amplifying the already potent mix of anthropogenic pressures behind what is increasingly called Earth's sixth mass extinction.

Extinction is not new. Ninety-nine per cent of all species that have ever lived on Earth have gone extinct. In addition to the established five mass extinction events, mostly driven by massive changes in climate from volcanic activity and other geological forces, countless minor extinction events have occurred, and

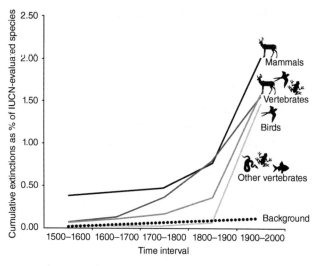

36. Cumulative extinctions of vertebrate species relative to background rate.

there are also background rates of extinction that are relatively continuous over the long term. To test whether humans are causing a mass extinction, it is therefore necessary to compare contemporary rates of extinction with the background rates of the past: the historical baseline for extinctions (Figure 36). Ecologist Stuart Pimm, palaeontologist Tony Barnosky, and others have demonstrated that current extinction rates for vertebrates, estimated in extinctions per million species per year, are now at least ten and potentially up to 1,000 times higher than the historical baseline and have increased dramatically in recent centuries.

It is hard to determine absolute rates of extinctions for a number of reasons. Extinction rates vary a great deal by taxonomic group. Vertebrates, especially mammals and birds, appear to be particularly vulnerable, while most plant taxa appear less so. Of Earth's estimated 5 to 10 million multicellular species, less than 2 million have been catalogued by scientists, so most extinctions probably occur before species are even known to exist. As of 2010, only 1,200 confirmed species were recorded as going extinct over the past 400 years. But confirming extinctions is much harder than confirming whether a species still exists. Imagine trying to prove that there are no bedbugs in Tokyo—much harder than confirming their existence. More concerning still is the prospect of extinction debt. Many populations have already become so small that narrowed gene pools and other restrictions on reproduction have made their future extinctions inevitable. For long-lived species, like trees, declining rates of reproduction may already have doomed some species even while their populations remain substantial. The American chestnut is an example—its ancient stumps, laid low by a European fungal disease, still sprout, only to die back again and again without ever setting seed.

Earth's sixth mass extinction has not yet arrived. Nevertheless, human societies are accelerating extinction rates well beyond their historical baselines, especially for vertebrates. Massive overfishing by factory ships is rapidly reshaping the biodiversity and food chains of entire oceans, defaunating the marine realm in much the same way as our ancestors defaunated Earth's land. If these rates of species loss are not curtailed, Earth's six mass extinction and a biosphere drastically reduced in biodiversity will come to define the ecology of the Anthropocene.

Homogocene

In 1958, Charles Elton published *The Ecology of Invasions by Animals and Plants*, calling attention to 'one of the great historical convulsions in the world's fauna and flora'. Massive losses of biodiversity were only the beginning. By transporting species around the world, humans were breaking down geographic barriers that had constrained the evolution of species for millions of years. By the 1980s, Gordon Orians and other ecologists were calling this global mixing of species the start of a new era, the 'Homogocene'. In becoming a global species, humans were taking the rest of the biosphere with them.

Humans have been introducing species to new areas at least since the late Pleistocene, when hunter-gatherers began propagating species they favoured. Still, the Homogocene most likely began in earnest only as human societies and their networks of long-distance trade expanded with the rise of agriculture and accelerated with the Columbian Exchange and the rise of global supply chains. Contemporary patterns of species introductions reflect this history, with the early industrialized trading nations of the north experiencing some of the greatest numbers, followed by areas industrializing later.

Introduced 'alien' or 'exotic' species are concerning because some have demonstrated the ability to outcompete, overconsume, and otherwise threaten the survival of native species, reshaping the biotic communities and ecosystems they invade in dramatic ways. For example, Kudzu, a vine from Asia, was introduced intentionally into North America as an ornamental plant and livestock feed. Within a few decades, Kudzu was known as 'the vine that ate the South', covering forests and causing annual damages greater than US$100 million per year. Kudzu is just one of many thousands identified as 'invasive alien species' and more than 500 of these have become problems around the world. Many common pests and diseases of crops, livestock, and wildlife are introduced species, causing damages estimated to exceed US$100 billion annually, and are held responsible for the extinction of nearly 40 per cent of all animals for which the cause is known.

Not all introduced species cause such harm, however. Many remain in the background or are even welcomed. In Europe, for example, species established outside their native range before 1492 are distinguished as 'archaeophytes' and considered 'more native' than later 'neophyte' arrivals even when known to be introduced by the Romans or others. European earthworms now predominate in North America over rare native earthworms, yet few consider them unwelcome despite their widespread transformation of soils and entire ecosystems. The concept of a stable native range also has little meaning outside the Tropics, and maybe not even there. For millions of years, species of the Temperate Zone have migrated up and down the continents to keep up with the glacial/interglacial cycles. With climate changing faster than ever, species must move to survive. In the Temperate Zone at least, definitions of native vs invader are challenged by a climate changing so rapidly that staying in one place is a recipe for extinction.

Human redistribution of species has already left a clear stratigraphic record of transformative ecological change across Earth. Yet the stratigraphic signatures of these changes, in the form of novel assemblages of species recorded in lake cores and other material deposits, may also be some of the most complex and diachronous of all markers of anthropogenic global change. The Homogocene is certainly here, but it is also here, there, and everywhere; a confusing mix of different changes at different times that offers many markers, but no single coherent signal to mark a lower boundary for the Anthropocene. It has also made conservation and restoration infinitely more challenging.

Shifting baselines

The classic approach to conservation and restoration has been to maintain or restore populations, environments, and habitats to their 'natural' states, defined in terms of a historical reference condition, or baseline. Assuming that a historical baseline can be established using palaeoecological or other historical evidence, two challenges remain. The first is selecting the appropriate baseline, and the second is managing ecosystems to stay within or return to this historical state. Both efforts are challenged by long-term anthropogenic ecological change.

In North America and Australia, for example, restoration and conservation efforts long embraced the 'Pristine Myth', defining 'natural' historical baselines as the state existing prior to first contact with Europeans; the environmental transformations sustained by native peoples were ignored. Yet this is but one of many possible 'natural' baselines. Were conditions more natural before the late Pleistocene megafauna extinctions? In between the various comings and goings of different human societies over millennia? By necessity, choosing a specific historical baseline among the many possibilities is more a matter of values than of science.

From a practical point of view, anthropogenic pressures have made it nearly impossible to restore and maintain historical baseline conditions in many if not most areas of the world. Biotic communities are being transformed by species losses while inundated by species invasions at the same time. The effect has often been a net gain in the total number of species within a landscape or region, but the newcomers are generally common weeds, pests, and other invaders—increasing both biodiversity and biotic homogenization. At the global scale, species are nevertheless lost to extinction while the Homogocene continues. Increasingly rapid changes in climate, soils, and other abiotic environmental conditions are only adding to and interacting with these changes in biota. For example, as a result of irrigation management, salt has accumulated in some South Australia soils, sustaining invasions of salt-tolerant non-native plants, some also adapted to warmer temperatures, yet reducing biodiversity overall.

Is it possible to sustain a historical reference state when both biotic communities and abiotic environments have shifted so far beyond their historic range of variability? Under such novel conditions, restoration ecologist Richard Hobbs and others have proposed that adhering to historical baselines may hinder conservation and restoration efforts more than it helps them. Hybrid ecosystems—part historical, part novel—might effectively be restored to their historical states. But traditional restoration is unlikely to succeed and too costly to consider in 'novel ecosystems' where biotic and abiotic conditions have shifted too far beyond historical levels.

Rambunctious garden

In the Anthropocene, baselines for conservation and restoration are shifting baselines, shaped by the changing values and anthropogenic ecological conditions created and sustained by human societies. What is the meaning of natural habitat or natural ecosystem when communities of plants and animals, and their relationships with each other and their environments, have all been transformed by prior histories of human social change? What does it mean to be a native species in an agricultural landscape or a city, where engineered soils, managed vegetation, excess nutrients, pollution, and other human-altered conditions are the norm—not a disturbance?

Perhaps you have marvelled at trees growing out of an abandoned building (probably *Ailanthus altissima* of *A Tree Grows in Brooklyn*), weeds growing up through the sidewalk, or even peregrine falcons hunting rats in the city. Species are learning to live in human environments, and some are getting very good at it. Along these lines, bird species with bigger brains, like crows and ravens, have been found to do better in complex human environments like cities. There is even evidence that introductions of alien species are speeding up the evolution of new species. Ecologist Chris Thomas has shown that in Britain, European rhododendrons have hybridized with their North American relatives to generate new wild populations, and a hybrid of two species of fruit fly has evolved to colonize invading honeysuckles in North America.

Most importantly, human societies are actively bringing back and learning to live with species they once killed off with impunity—witness the return of wolves to their ancient hunting grounds in Europe and black bears, mountain lions, and coyotes across the USA. Life still thrives in what writer Emma Marris has called the 'rambunctious garden' of the Anthropocene, in which novel ecosystems form the new wild. In an increasingly anthropogenic biosphere, new relationships are forming. Societies, people, wildlife, and entire ecosystems are co-evolving and co-creating new forms of nature in addition to conserving and restoring those that came before.

Social-ecological systems

Ecologists are increasingly probing the causes and consequences of anthropogenic ecological change and developing new paradigms that embrace the coupling of human and natural systems. In the 1950s, ecologist Eugene Odum highlighted human dependence on ecosystems in the textbook that helped make ecology a household word in the 1960s and 1970s. He also studied 'old field succession', the recovery of vegetation on abandoned farmland. Rachel Carson brought an ecologist's understanding of the widespread consequences of industrial chemicals to the public with *Silent Spring* in 1962. Ecosystem research in the Hubbard Brook watershed led to the discovery of acid rain in the 1970s. And in 1986, Peter Vitousek went global with his estimate that humans were 'appropriating' nearly 40 per cent of Earth's terrestrial photosynthesis by harvesting forests and using land for agriculture. Before Crutzen's Anthropocene, Vitousek made the case for an Earth reshaped by humanity in his classic 1997 paper in *Science*, 'Human Domination of Earth's Ecosystems'.

By the late 1970s, ecologists were incorporating humans more and more deeply into their research and partnering with social scientists to investigate the coupling of social and ecological processes. Urban ecology, industrial ecology, and agroecology emerged as distinct sub-disciplines, and landscape ecologists, conservation biologists, and other applied ecological disciplines incorporated managed ecosystems into their work. In the 1990s, Carl Folke developed a popular framework for 'social-ecological systems', accelerating collaborations among ecologists and social scientists to solve real world problems involving environmental management and social change (Figure 37). Ecological economics (ecology first), environmental economics (economics first), and other disciplines allied with ecology introduced new tools to address environmental management challenges, including the recognition, measurement, and management of 'ecosystem services', the societal benefits provided by ecosystems, such as pollination, clean drinking water, and recreational opportunities.

Ecologists have also scaled up their work, bringing models of an 'active biosphere' to Earth system simulations in the 1990s, a major advance over models that assumed vegetation stayed put even in the face of major climate change. Imagine trees forced to live in deserts, and a global carbon cycle governed by physics alone.

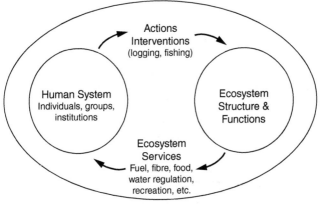

Social-Ecological System

37. Diagram illustrating the coupled interactions of social-ecological systems.

Ecologists, economists, geographers, and others are now developing approaches to observing, understanding, and modelling a global ecology actively shaped by human societies, including changes in human use of land—the largest driver of biodiversity change and anthropogenic carbon emissions to the atmosphere until 1950.

Anthropogenic biosphere

The global patterns of Earth's terrestrial ecosystems have long been shaped by climate, terrain, soils, and other abiotic environmental constraints to which species are adapted. Deserts are populated by plants adapted to dry environments, species living in tropical rainforests like it warm and wet, and patterns of vegetation change from the bottom to the top of high mountains. This global environmental patterning of life was first described by Alexander von Humboldt in the early 1800s, prompting the development of the field of biogeography in the early 1800s. By the mid-1930s, ecologists were describing these global patterns as 'biomes', or ecosystem patterns at the global scale, one step below the largest scale of all, the biosphere.

As ecologists came to grips with an increasingly anthropogenic biosphere, efforts were made to understand the global patterns of ecology shaped by humans. By the 1990s, satellite remote sensing was producing the first global maps of vegetation cover, including anthropogenic covers like crops and artificial surfaces—even the glow of night-time lights. In 2002, ecologist Eric Sanderson combined these data with maps of roads and human population density to map an index of increasing human influence, leaving wildlands as areas without them. The global patterns of human transformation of ecology, estimated to cover more than 80 per cent of Earth's land, were becoming clearer, yet they still appeared as no more than a disturbance of an otherwise natural world.

With most of the terrestrial biosphere reshaped by humans, the need for a rich understanding of the global ecological patterns produced by human interactions with ecosystems became clear. In 2007, I worked together with geographer Navin Ramankutty to rectify this by integrating data on human populations, land use for crops and pastures, and vegetation cover to map Earth's anthropogenic biomes, calling these anthromes (Figure 38). Our data showed that in year 2000, more than 75 per cent of the terrestrial biosphere had been transformed into anthromes, including urban areas and other dense settlements (around 1 per cent of Earth's ice-free land), agricultural villages (6 per cent), croplands (16 per cent), rangelands (32 per cent), and seminatural lands with only minor human populations and land use (20 per cent), leaving wildlands without human populations or land use in less than one-quarter of the terrestrial biosphere. In later work, we showed that significant areas of anthromes first emerged about 8,000 years ago, and covered more than half of the terrestrial biosphere between 500 and 2,000 years ago, depending on the historical data we used, but this was mostly in the form of seminatural lands. Only in the past century has more than half of the terrestrial biosphere been transformed into the most intensively used urban, village, cropland, and rangeland anthromes.

One key finding in assessments of human transformation of the terrestrial biosphere is that even within the most densely populated and intensively used anthromes, including cities and villages, substantial areas are left without intensive use, sometimes intentionally as parks, but mostly because both farmers and developers tend to avoid mountains, hills, and other environments less suitable for agriculture and infrastructure. As a result, anthrome landscapes are generally mosaics of used lands interspersed with less used, recovering, and remnant ecosystems transformed by being broken up and embedded within used landscapes, subjected to hunting, fuel gathering, species invasions, pollution, and other human pressures. Even though only 40 per cent of Earth's land is used directly for crops, pastures, and settlements, this has transformed another 35 per cent into novel ecosystems with biotic

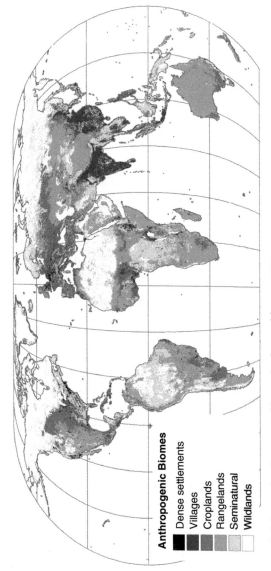

38. Global map of anthropogenic biomes (anthromes) in year 2000.

Anthropogenic Biomes

- Dense settlements
- Villages
- Croplands
- Rangelands
- Seminatural
- Wildlands

communities and ecological processes long departed from any 'natural' historical baselines.

Human societies are far more than a disturbance to an otherwise natural world. Human social systems have emerged as a planetary force within the Earth system—an anthroposphere that is actively shaping and sustaining an anthropogenic biosphere. Human social networks are now woven globally into the web of life. Decisions made in one place can change ecology on the other side of the planet and even globally; human and natural systems are globally 'telecoupled'. As humans continue constructing their niche across the planet, Earth is functioning more and more like a social-ecological system with a social metabolism geared toward sustaining increasingly wealthy and demanding human populations. Already, more than 90 per cent of Earth's total mammal biomass is composed of humans and domesticated animals. How far can this go? Are there no limits to how many people and how much transformation Earth's ecology can handle?

Limits to growth

Long before Thomas Malthus published his *Essay on the Principle of Population* in 1798, the question 'how many people can Earth support?' was asked and answered many times. For example, Antoni van Leeuwenhoek computed this figure to be 13.4 billion in 1679. Nevertheless, ever since Darwin used Malthus' dictum that populations are limited by scarce resources to explain his theory of natural selection, this concept has been central to scientific debates about the planetary limits of human populations. In the 1920s, ecologists formalized this as 'carrying capacity' (K): the environmental limits to a population's growth in a given environment. When populations grew beyond their carrying capacity, it was argued, a crash was imminent.

Concern over limits to human carrying capacity on Earth came to a head in 1968 with Stanford ecologist Paul Ehrlich's book

The Population Bomb, which predicted that 'hundreds of millions would starve to death' in the 1970s from overpopulation. In 1972, an influential book, *The Limits to Growth*, used early computer simulations to explore the grave consequences for the 'natural ecological balance of the earth' when populations grew beyond a 'global equilibrium'. In 1994, Ehrlich stated that 'the present population of 5.5 billion...has clearly exceeded the capacity of Earth to sustain it'. Paul Ehrlich has made major contributions to the science of ecology, but the famines he predicted have yet to occur.

Earth's current population of more than 7 billion is much better fed, healthier, and living longer than at any time in human history. Rates of population growth have slowed dramatically since the 1970s and are continuing to decline (Figure 39), mostly as the result of the 'demographic transition', in which more urban,

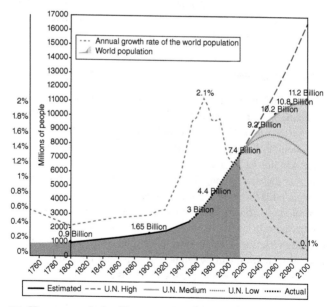

39. **Human population history, projections, and growth rates, 1800 to 2100.**

better-educated populations tend to have much smaller families. Earth's human populations are continuing to urbanize and population growth rates are continuing to drop. It is possible that human populations might reach 16 billion by 2100 and continue growing, but the mainstream prediction of demographers is that populations will level off at about 11 billion in 2100.

Planetary boundaries

Though population growth is slowing, human demands for food, water, energy, and other environmental resources are continuing to grow, as wealthier populations make greater demands on Earth's resources. For example, one environmental group has claimed that humans now use the equivalent of 1.6 Earth's worth of resources to sustain themselves—an unsustainable 'overshoot' of Earth's biological carrying capacity. Moreover, many scientists and others are concerned that even current levels of population and resource demand may be harming Earth's 'life support systems' in ways that might prove catastrophic in the future. Accelerating global climate change is just one of many potential catastrophes.

In 2009, a group of scientists, including Will Steffen and Hans Joachim Schellnhuber, identified nine such Earth system changes in *Nature*, highlighting their 'planetary boundaries', which, 'if crossed, could generate unacceptable environmental change' (Figure 40). Drawing on the concept of tipping points in the Earth system, the crossing of which might force Earth out of its 'stable, Holocene-like state', the planetary boundaries framework raises the possibility of catastrophic changes if the Earth system is pushed too far.

Recently, this framework and later versions of it have been called into question both scientifically and as a rubric for environmental governance. Except for climate change, ozone depletion, and

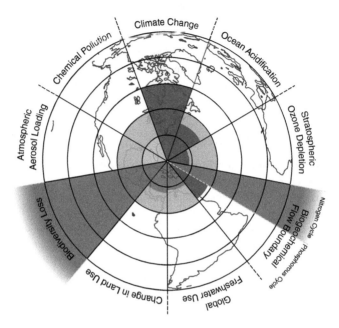

40. Planetary Boundaries. For nine global environmental variables, light grey area highlights 'safe limits', with dark grey shading indicating that these limits have been exceeded (biodiversity loss, climate change, and changes in the nitrogen cycle).

ocean acidification (in Chapter 8), scientific evidence for tipping points in the Earth system is very limited. Many of the Earth system changes in the framework are cumulative, resulting from local and regional changes that add up, and not the kinds of systematic planetary changes, like the atmospheric build-up of greenhouse gases, known to produce tipping points. From a policy perspective, setting safe levels for Earth changes is risky, especially if such levels are not strongly supported by scientific knowledge, because these suggest that below some threshold nothing serious will happen, but beyond them, change is inevitable. Holding such a view risks breeding complacency on one side and hopelessness on the other. Both are misplaced: to lose even one species is more than we should accept lightly. The same holds for local habitats. Even so, the call to avoid transforming Earth in ways that cause serious harm to both humans and non-human nature has helped raise these serious scientific concerns to the global level.

Yet the question remains: if human societies are now operating as a global force that is transforming Earth to the detriment of both humanity and non-human nature, what, if anything is to be done about it? Who is responsible? Who shall act?

Chapter 7
Politikos

In an influential 2009 paper entitled 'The Climate of History: Four Theses', historian Dipesh Chakrabarty asked, 'is the Anthropocene a critique of the narratives of freedom?' The asking of such a question exemplifies just how far the concept of the Anthropocene has travelled beyond its origins in the natural sciences. In the two decades since Crutzen first proposed it, the Anthropocene has inspired a barrage of socially relevant questions, stoked intense debates, and served as a muse for artists and designers.

While stratigraphers work to define a golden spike, others are questioning the meaning and implications of a new age of humans. The politics of inequality, environmental ethics, and the challenges of responsible action under conditions of potentially catastrophic global change have all been connected with the Anthropocene proposal. Even Stan Finney, former chair of the International Commission of Stratigraphy, has questioned whether the Anthropocene is more of a political statement than a scientific imperative.

Hubris

For some philosophers, conservationists, and even geologists, the act of designating a human epoch says more about human hubris and anthropocentrism than it does about science. Who are 'we' to name a new interval of geologic time after ourselves, and why are we doing it? The 'Atomic Age', the Homogocene, the 'Carbocene' (an 'Age of Fossil Fuels'); all might amply describe our time. Why choose a name that puts our species in the foreground?

Sociologist Eileen Crist and others have argued that recognizing an 'age of human dominion' only serves to justify human ownership and destruction of nature, paving the way towards grander projects for transforming nature even further. Even geologist Stan Finney was concerned that efforts to make the Anthropocene 'official' were intended to create a political 'mindset' to 'take control of the current transformation'. Towards this end, some have called attention to Paul Crutzen's long-standing interest in geoengineering—the intentional manipulation of the Earth system to control anthropogenic climate change. By formally recognizing an Earth transformed by humans, the Anthropocene might, they argue, serve as the political equivalent of stating 'anything goes' and dismissing all efforts to limit human transformation of Earth as foregone and impossible. Ecologist Edward O. Wilson has labelled advocates of such thinking 'Anthropocenists'—though it is not clear who precisely he is referring to.

Similarly, the prospect of a formal Anthropocene Epoch has troubled some conservationists, who interpret this as marking all of nature as 'touched by humans', seemingly leaving no nature left to conserve. Opposing the Anthropocene, they point out that declaring Earth's ecology entirely transformed by humans exaggerates the scope of human alterations while cultivating 'hopelessness in those dedicated to conservation'. Others, arguing that significant 'naturalness' would remain on an Anthropocene Earth, suggest that rarity might only enhance the value of nature conservation. Either way, multiple arguments against recognizing the Anthropocene have been raised by those concerned about the negative consequences of reifying an 'age of humans' and a nature profoundly transformed by humanity.

Epochalypse

Perhaps the most popular interpretation of the Anthropocene, though, is as a catastrophic, human-induced shift in Earth's functioning as a system. In this view, recognizing the Anthropocene is the same as acknowledging the serious global consequences of climate change, mass extinctions, and other anthropogenic environmental changes. In the words of philosopher Clive Hamilton, 'Earth has now crossed a point of no return,' a 'rupture' in Earth's functioning that 'should frighten us'. Or as geographer Erik Swyngedouw has written, 'The Anthropocene is just another name for insisting on Nature's death.' Failing to formally recognize the Anthropocene or interpreting it in some other way is therefore equivalent to denying the significance of global environmental change.

Earth system scientists have indeed used the Anthropocene as a kind of shorthand for human transformation of Earth's functioning as a system. Yet the Anthropocene itself is a synthesis of existing evidence, and not a new source of evidence for these changes or their consequences. For scientists in general, the evidence that humans are causing potentially catastrophic changes to Earth's functioning as a system is rich, multifaceted, detailed, and robust—the product of decades of research. There is no need for an Anthropocene epoch to understand or recognize these changes. Indeed, a growing number of Earth scientists, including Stan Finney, are increasingly concerned that the push to recognize an Anthropocene epoch might even be diverting scientific efforts away from more important goals, like better understanding and addressing the specific challenges of global climate change or mass extinction. In the words of geologist James Scourse, 'while the anthropocenists rearrange the deck chairs, other scientists are getting on with the business of trying to understand, and do something about, the crisis we face'.

Equating the Anthropocene with anthropogenic global environmental change raises yet another concern—the possibility that it might not be accepted as a new epoch of geologic time. What would be the message then?

Despite overwhelming scientific evidence, the broader public in some nations remains divided over the seriousness of anthropogenic climate change, accelerating extinctions, and other environmental changes with serious global consequences. Would scientific recognition of the Anthropocene change public perceptions and actions to better avoid or adapt to these changes? As with the Anthropocene itself, the jury is still out.

Colliding histories

Long before C. P. Snow criticized the 'two cultures' separating the sciences and the humanities, the study of human and natural history were separate. While historians did acknowledge that humans could alter natural environments and that the course of human history could be altered by natural disasters like floods and droughts, any causal interconnections between the two were usually ignored. Human societies occupied centre stage, and natural environments, the background.

In the 'Climate of History', Dipesh Chakrabarty argued that with anthropogenic global climate change, the separation of human and natural histories was over for good. By changing climate, humans became 'geophysical agents' and a 'force of nature' qualitatively distinct from the past, when humans interacted with nature only as 'biological agents'. With climate change, Chakrabarty argued, 'The geologic now of the Anthropocene has become entangled with the now of human history.' And with this entanglement, nature and society became one.

While Chakrabarty's paper has stimulated intense discussions across the humanities, he was not the first to recognize the artificial separation of nature and society. Anthropologists, critical social theorists, and environmental historians like John McNeill have been linking human and natural histories for decades. An entire field of environmental humanities has now been built upon this very foundation. Chakrabarty's recognition of a substantial difference between geological and biological agency has also been criticized; changing the biosphere alters Earth's atmosphere, climate, and other processes just as surely as burning fossil fuels.

Most concerning for many, including Chakrabarty himself, has been the implied need to understand the Anthropocene as a 'species history' of the 'Anthropos', lumping every person on Earth into a single undifferentiated mass—the exact opposite of what the humanities have been doing at least since the 1960s. Chakrabarty even accepted, reluctantly, the mainstream Anthropocene narratives of the natural sciences, in which humans as a species would need to be guided by Enlightenment values of rationality in order to address the unprecedented social and environmental challenges of the Anthropocene. Unsurprisingly, these 'enlightened species' narratives have been challenged, leading to new narratives and names for societies' collision with nature.

Whose Anthropocene?

No other species has ever transformed Earth as much as we have (the Great Oxygenation involved dozens if not thousands of cyanobacterial species). Even in small numbers, human hunter-gatherers caused widespread extinctions and biospheric changes that may have altered Earth's climate. Yet anatomically modern humans behaved no differently from their ancestors for tens of thousands of years, and these early environmental changes pale in comparison with today's.

It should be clear that there is no one 'human' way of transforming Earth. Different people use and transform environments differently, producing different consequences, and different people experience these consequences differently. Have you ever burned coal or grown wheat? Unlikely. Yet these have almost certainly been done for you. Your home, your consumption of resources, and your exposure to environmental hazards—all of these are a function of your society and your role in it. The human way of living on Earth, our ecological niche, is shaped far more by our societies than by our biology. And different societies use and transform environments very differently.

Even now, when human societies have never been more interconnected, there are striking differences in societal use of resources. Carbon dioxide emissions from Earth's three most populous nations are illustrative (Figure 41). In 2014, China, with 1.4 billion people, emitted 10.5 billion tonnes (Gt) of CO_2, or about 7.6 tonnes per capita. The USA's 320 million people emitted 5.3 Gt; 16.7 tonnes per capita. India's 1.3 billion people produced 2.3 Gt of CO_2, or about 1.6 tonnes per capita. Though China emits twice as much CO_2 as the USA, the average Chinese person produces half the emissions that an average US person does. An average Indian person produces only one-tenth as much, and 100 people in some African nations emit less than one average US citizen. And differences within nations can be just as stark. For example, wealthy urban dwellers can emit ten times or more per person than the rural poor. And Earth's poorest billion people emit almost no fossil carbon at all.

Is it correct to say that *Homo sapiens as a whole* is causing rapid global climate change? Clearly not. Wealthy nations and wealthy people use vastly more energy and emit far more carbon dioxide than the poor. Travelling by personal car and jet aeroplane, which most people on Earth have never done, are some of the most energy intensive things a human can do. And until very recently, virtually all of this energy came from cheap, abundant, fossil fuels. The consequences have been wealthy, carbon-intensive lifestyles for some, and a carbon-filled atmosphere for all.

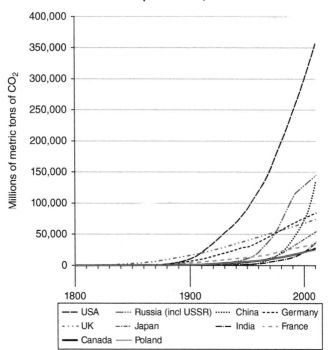

**Cumulated CO$_2$ equivalent emissions
Top 10 nations, 1800–2010**

Millions of metric tons of CO$_2$

— — USA	— · · · Russia (incl USSR)	· · · · · · China	- - - - Germany
· · · · UK	—·· – Japan	—·— India	– – France
—— Canada	—— Poland		

41. **Cumulative carbon emissions, 1800 to 2010.**

Global climate change will not end without ending carbon
emissions from fossil fuels. But without other cheap sources of
energy, fossil fuels might remain the pathway to wealth. Powered
by a soaring industrial economy, China lifted itself from the ranks
of the poor to the economic powerhouse of the world, racing past
the USA to become Earth's largest annual emitter of carbon dioxide
around 2005. Based on current emissions, China is easily targeted
as the main producer of global climate change. Yet this simple
assessment hides deeper inequalities. China only began burning

large amounts of fossil fuels in the 1980s as it fuelled its way to industrial wealth. The USA reached similar levels a whole century earlier, and the UK decades before that. Even now, China has a long way to go before it can match the USA's total emissions since 1850 (Figure 41). Moreover, China's production for export to the rest of the world accounts for one-third of its total emissions. Even China's emissions are not China's alone.

Capitalocene

Naming an epoch after humans appears to blame people, in general, for transforming Earth. But people have never transformed Earth equally. The wealthiest humans in the wealthiest societies are the main cause of rapid global climate change. To blame everyone equally is like blaming the bank for being robbed. Billions of people have never used cheap fossil energy to lighten their loads.

Blaming humans as a whole also avoids the most important question of all. Where did all of this inequality come from? Anthropogenic environmental change is a social process. A single person may flip a light switch, but it takes an entire society to keep the lights on. Inequalities in human transformation of environments are merely a reflection of inequalities within and among societies resulting from socio-political and economic processes.

One of the most talked about alternatives to the Anthropocene puts the blame squarely on a single social transformation. For human ecologist Andreas Malm, geographer Jason Moore, and anthropologist Alf Hornborg, calling our time the Anthropocene diverts attention from the real culprit behind anthropogenic environmental change.

In the words of Jason Moore in 2014, the Capitalocene began with 'a turning point in the history of humanity's relation with the rest of nature, greater than any watershed since the rise of agriculture and the first cities—*and in relational terms, greater than the rise of the steam engine*'. Capitalism, not industrialization, caused Earth's transformation by producing massive social inequalities that supported 'audacious strategies of global conquest, endless commodification, and relentless rationalization'.

Malm went further, claiming, 'uneven distribution is a condition for the very existence of modern, fossil-fuel technology'. And activist Naomi Klein put it even more starkly:

> Fossil fuels require sacrifice zones: they always have. And you can't have a system built on sacrificial places and sacrificial people unless intellectual theories that justify their sacrifice exist and persist: from Manifest Destiny to Terra Nullius to Orientalism... In this way, the systems that certain humans created, and other humans powerfully resisted, are completely let off the hook. Capitalism, colonialism, patriarchy—those sorts of system.

On this basis, advocates for the Capitalocene have criticized Anthropocene narratives from the natural sciences as 'ahistorical and apolitical'. Portraying global environmental change as the product of an undifferentiated humanity hides the political realities behind these changes, including who benefits and who loses. In *The Shock of the Anthropocene*, published in 2016, historians Christophe Bonneuil and Jean-Baptiste Fressoz go further, characterizing this disregard of politics as far more than just a naive oversight. In their view, the elites responsible for harming environments have always been aware of their negative consequences and have always worked to obscure them from public view.

While it is difficult to establish just how much anthropogenic environmental change may have been facilitated by covering up the consequences, such cover-ups have certainly occurred. For example, historian of science and AWG member Naomi

Oreskes has documented how major corporations in the fossil fuel industry concealed their early knowledge of anthropogenic climate change and funded campaigns to sow doubt about its scientific underpinnings.

Even more troubling for those concerned about a Capitalocene conspiracy are Anthropocene narratives in which a global human 'awakening' to the environmental hazards of the Anthropocene produces new regimes of global environmental governance controlled by an elite-serving technocracy. Such narratives not only whitewash the environmental wrongdoings of a hegemonic capitalist elite, they amount to a political strategy in themselves. By recognizing the Capitalocene and rejecting a technocratic 'species narrative' of the Anthropocene, a space for more nuanced and politically aware strategies is emerging in response to the unprecedented global environmental challenges of the current time.

Governance

'The Anthropocene has to be named before people can try to take responsibility for it,' wrote law professor Jedediah Purdy in his 2015 book *After Nature: A Politics for the Anthropocene.* Still, Purdy, like most others, also highlighted that there remains no clear-cut political constituency or governance infrastructure prepared to take on the overwhelmingly complex and wicked social-environmental challenges of the Anthropocene.

To call Anthropocene challenges 'wicked' is not to call them evil (though some might well be) but to highlight that they are perfect examples of what policymakers call 'wicked problems', characterized by the absence of agreed-upon solutions, the tendency of solutions to yield additional problems, for solutions to generate both winners and losers, and the difficulties of even defining what the problems are. Two simple examples are habitat loss and climate change. In both cases, causing these problems provides clear benefits to some (food production, cheap energy) and yields environmental harm that is both difficult to compute and affects various groups of people differently. Agricultural land can be taken out of production to provide habitats for other species, but where will food then be grown? Is it better to solve fossil fuel emissions by technologies that remove the carbon, like carbon capture and storage (CCS)? Or is it better to invest in alternative sources of energy, like solar or nuclear, that would displace fossil fuels altogether? Or some mix of solutions? And this is only the tip of the iceberg. Who wins, who loses, who bears the costs, and who decides? In the Anthropocene, all of these issues remain firmly on the table.

On the surface, it would seem that global environmental governance would be necessary to tackle global environmental problems. Yet efforts to address global climate change through frameworks for international governance have so far generated more failures than solutions. While the Montreal Protocol and its follow-up agreements have largely succeeded in protecting Earth's ozone layer, a similar effort to cap carbon dioxide emissions in the 1990s, the Kyoto Protocol, is a case study in failed environmental governance. The most recent international framework for climate change prevention, the Paris Agreement of 2016, did achieve universal international agreement that humans are causing global climate change for the first time. Yet it is also probably the weakest international framework so far, including no mandatory actions or binding commitments to reduce greenhouse gas emissions.

Efforts to solve global environmental problems continue to focus on international laws and agreements, including a newly proposed framework of 'legal boundaries to stay within planetary boundaries' and efforts to expand the International Law of the Sea. Nevertheless, policy scientist Frank Biermann, chair of the Earth System Governance Project, and others, have argued that the Anthropocene calls for novel governance strategies that recognize that recent rates, scales, and processes of environmental change are both unprecedented and interconnected in surprising ways, while also engaging directly with the complex inequalities that characterize both human populations and the environmental changes they create.

Is global governance the key to solving global environmental problems, or are other actors, like corporations, non-governmental organizations, and city governments, critically important? Must global environmental governance be democratic—one person one vote—or ruled by nations or other institutions? How to address policies that solve problems in one sector of society while creating problems in another, as when fertilizer subsidies boost food production in one region while shutting down coastal fisheries in another? Can the needs of future generations be included in policies and governance today? And what of the rights of non-humans on a planet increasingly transformed to serve humans?

Chthulucene

Perhaps the most challenging counternarrative to the Anthropocene, even to pronounce, is the 'Chthulucene', introduced by feminist theorist and philosopher of science Donna Haraway in 2014 and published in journal form in 'Anthropocene, Capitalocene, Plantationocene, Chthulucene: Making Kin'. For Haraway and a broad community of fellow travellers in the humanities and social sciences, the Anthropocene's focus on the human is a problem in itself. By reproducing an imagined world under human control, Anthropocene 'species thinking' demands to be destabilized through confrontations with 'other worlds of thinking' that decentre and entangle humans within complex webs of social and ecological processes within which non-humans play key roles.

In embracing the Capitalocene as a useful species-decentring challenge to the Anthropocene, Haraway also calls out the Plantationocene, characterized, like the Orbis spike, by

> the devastating transformation of diverse kinds of human-tended farms, pastures, and forests into extractive and enclosed plantations, relying on slave labor and other forms of exploited, alienated, and usually spatially transported labor.

With the Chthulucene, Haraway takes her argument much further by using the mythical, cosmic, multi-tentacled god-like alien of writer H. P. Lovecraft ('Cthulhu') to symbolize the impenetrable entanglement and interconnectedness of what only appear to be individual beings. To build this vision, Haraway draws on recent scientific evidence that individuals of most, if not all, species are actually functioning assemblages of multiple species. For example, human individuals contain more microbial than human cells, mostly within the biodiverse 'microbiome' of our digestive systems. For Haraway, individuality is just an illusion.

Imagining an Anthropocene world of human control is to embrace the 'extinctionist' paradigm that transformed Earth in the first place, Haraway argues. Reinventing the human as an entangled imaginary embedded within a broader world of co-dependent multispecies assemblages can reverse the destructive narratives that have justified and guided human transformation of Earth. To facilitate such processes of human reinvention, Haraway coins a slogan, 'make kin, not babies!', embracing as kin all of Earth's 'critters', including all of Earth's humans, through our common ancestry, bringing all of life together within the biosphere.

While Haraway intentionally distances her narratives from scientific language, they still seemingly overlap with the systems-thinking of Earth system science, in which all of Earth's organisms are functionally interconnected with each other and with Earth's abiotic environments through biogeochemical cycles and flows of energy, leading to the emergence of surprising non-linear dynamics.

The Chthulucene's deep social-ecological perspective also highlights one of the Anthropocene's most challenging ethical questions: whether humans have any right to change Earth in the first place. By colliding the Anthropocene with the Chthulucene, Haraway comes to the assistance of a broader community of 'posthumanist' social movements that reject 'interspecies hierarchies' towards new forms of bioethics including animal liberation, an 'Animal's Agenda', and the recognition of an 'intrinsic value of nature' that goes beyond human systems of valuation. As environmental humanities professor Ursula Heise has written, it is time for 'a more-than-human democracy that turns the idea of the Anthropocene away from its uniquely human focus'. And in so many ways, these new movements are only rediscovering the cultural values, perspectives, and narratives that many non-Western societies have always held.

Reflections

By stimulating awareness that the world we inhabit is increasingly of our own creation, the Anthropocene is also emerging as a 'reflexive time period', a time in which humans reimagine what it means to be human. Such reflections have inspired a broad ferment of ideas and artistic expressions far beyond any imagined by the Anthropocene's first proposers. Workshops, conferences, and other gatherings are popping up both inside and outside the academy, broadening the community of scholars, thinkers, and creators engaging in the reimagining of humanity and nature.

One of the broadest efforts to stimulate new thinking around the Anthropocene has come through the Anthropocene Project of the Haus der Kulturen der Welt in Berlin (HKW; 'House of World Cultures'), beginning in 2013. Generously funded by the German government, the HKW has curated multiple exhibits and 'gatherings' of scholars, artists, and performers invited from all over the world to conduct 'basic cultural research' together, taking as the 'core premise of the Anthropocene thesis' that 'Our notion of nature is now out of date. Humanity forms nature.'

As an invited participant in the initial 'Opening' gathering, I was astonished by the broad array of Anthropocene interpretations on offer. My own invited presentation on an Anthropocene 'thing' focused on a 'rock' formed of molten metal rubbish that I had found as a teenager, and a broken brick with its producer's name from a demolished Chinese city wall. Jan Zalasiewicz was there too, presenting a live cat, and both Dipesh Chakrabarty and Will Steffen gave keynotes. Engaged in a public discussion of the question 'Is the Anthropocene Beautiful?' with Emma Marris, I remember being disappointed that I had missed out on a performance by a naked man I was later told had transformed himself into a wild feline at the same time a genuine wild fox raced across the stage. The Anthropocene had become a very strange new world indeed. The HKW has also hosted an 'Anthropocene Campus' and produced an 'Anthropocene Curriculum', and is continuing to support work along these lines. I have a T-shirt from one of these with the text 'When Are We?' In a rare reversal of fortunes between arts and sciences, the HKW also hosted the inaugural scientific meeting of the AWG, which does not have funds of its own to do this.

The Anthropocene has appeared in books entitled 'Art in the Anthropocene', 'Architecture in the Anthropocene', 'The Birth of the Anthropocene', 'Adventures in the Anthropocene', 'Learning to Die in the Anthropocene', 'Love in the Anthropocene', and a volume of poetry titled 'The Misanthropocene'. There is an instrumental work, 'Deep Anthropocene', by musician Brian Eno, and a song, 'Anthrocene', by Nick Cave and the Bad Seeds. There is a popular heavy metal album entitled 'The Anthropocene Extinction' (by Cattle Decapitation), featuring artwork of apocalyptic post-industrial landscapes strewn with human corpses spewing plastic debris.

There are multiple documentary films on the Anthropocene and more on the way.

There appears to be a common thread across the Anthropocene's more creative interpretations: the Anthropocene as crisis. A crisis of nature, a crisis of humanity, a crisis of meaning, a crisis of knowledge, and above all, a crisis of action. The Anthropocene demands action.

Chapter 8
Prometheus

In 1999, Hans Joachim Schellnhuber asked the pivotal question of the Anthropocene. 'Why should Prometheus not hasten to Gaia's assistance?' If humans are indeed transforming Earth, what is to be done? Or more humbly—what can be done? Can humans help to bend Earth's trajectory towards better outcomes for both humanity and non-human nature?

The science is clear. Human well-being is generally improving at the same time that our societies are rapidly producing a hotter, more polluted, less biodiverse, and less predictable planet. The entire Earth system is being forced into a state with no analogue in its history, introducing the very real possibility of environmental changes so rapid and so powerful that even the most resourceful societies on Earth might not survive them. To continue along such a trajectory is to gamble with the very future of both human societies and the rest of life on Earth.

What's at stake, outside the domains of geology and stratigraphy, is a new account of our place in nature, our relationship with the rest of the planet. This narrative raises some hard questions, like, what exactly are we doing with our planet? Is this a story of senseless destruction or a story of awakening and redemption? It is clear we have only just begun to understand the many dimensions, variations, and alternatives that could play out in the future of the Anthropocene. Maybe the point, at this stage, is not which account to believe, but the need for many different Anthropocene narratives, to engage with the broadest range of human needs? A single account of our place on Earth has never been enough for most human societies.

Could recognizing the Anthropocene spark action towards better futures? For the editors of *Nature* remarking on the May 2011 meeting on the Anthropocene at the Geological Society of London, the answer was clearly yes: 'Official recognition for the Anthropocene would focus minds on the challenges to come.'

Driving the spike

Interest in the Anthropocene held steady after Crutzen introduced the term in 2000. But work in this area really got going after 2008 when geologists got involved and simply skyrocketed after 2011. Only a few dozen people attended the London meeting, but Zalasiewicz and colleagues had done their homework. The March issue of *National Geographic* featured the Anthropocene. Work by invited speakers had just been published in a special Anthropocene-themed issue of the *Philosophical Transactions of the Royal Society*, founded in 1665 (Newton and Darwin both published there). 'Welcome to the Anthropocene,' announced the cover of *The Economist*.

In the January 2016 issue of *Science* magazine, the AWG (including myself) presented scientific evidence supporting the Great Acceleration of the mid-20th century as the main scientific narrative explaining Earth's transition out of the Holocene Epoch and into the Anthropocene. Extinctions, deforestation, domestication, species invasions, agriculture, rice farming, anthropogenic soils, and even the Industrial Revolution were all examined and rejected as too diachronous to define a globally synchronous golden spike for the proposed new epoch of the GTS. The AWG also considered and rejected Lewis and Maslin's Orbis proposal based on the CO_2 dip in 1610, for reasons of relative size of the signal and difficulties in global correlation, in the same *Science* paper and subsequent publications.

At the August 2016 meeting of the International Geological Congress in South Africa, the results of a vote among the AWG's 35 members were presented, demonstrating near unanimous support for Anthropocene recognition, and only 3 votes against its formalization. A mid-20th-century start was also well supported, though 4 members did vote for a 'diachronous' beginning. Votes for specific stratigraphic markers varied, with 10 supporting plutonium fallout, 4 supporting radiocarbon, 3 votes for plastic, and 6 abstaining.

Outside the AWG, support among geologists has been mixed. Criticisms vary, but the most general concern, voiced by Stan Finney, Phil Gibbard (former chair of the ICS Subcommission on Quaternary Stratigraphy), William Ruddiman, Whitney Autin, John Holbrook, and James Scourse, amounts to strong doubts about the Anthropocene's utility to geological science. In the words of Scourse in 2016:

> There are plenty of ways of measuring time and establishing stratigraphies for the epoch when humans have progressively impacted the Earth system, such as measuring tree rings, radioisotopes introduced by atomic weapons testing, or counting annual layers in ice cores. We use these tools on a daily basis and have no need for the new term.

As I write, the AWG continues to winnow down potential GSSP candidates from among dozens found in lake sediments, peat bogs, glaciers, caves, and other stratified deposits. If all goes well, a formal Anthropocene GSSP proposal will be ready before the 2020 meeting of the International Geological Congress in India.

Going deeper

The Anthropocene continues to be controversial across the many scholarly communities that study social and environmental change, including not only archaeologists, anthropologists, sociologists, geographers, and environmental historians, but also ecologists and Earth scientists. A common concern is simply timing. Evidence for human transformation of Earth is abundant long before the 20th century and can be traced back into the late Pleistocene. But the most widespread concern relates to what archaeologist Andrew Bauer has called the 'Anthropocene divide'.

Stratigraphers divide geologic time into discrete intervals for purely pragmatic reasons, not because they believe that Earth's dynamics are not continuous. Nevertheless, efforts to advance scientific understanding of human transformation of Earth are inherently focused not on the identification of precise boundaries in time, but on the complex, continuous, socially differentiated, ecologically connected, and historically contingent processes by which humans have been producing this transformation over time. From this broad perspective, it is hard to see how dividing geologic time into two parts around 1950, as the AWG proposes to do, or even 7,000 years ago as some have argued, will help to advance scientific efforts to understand Earth's human transformation.

Archaeologist Karl Butzer called the Anthropocene an 'Evolving Paradigm'. Archaeologists Bruce Smith, Melinda Zeder, and Tod Braje have proposed that a combined 'Holocene/Anthropocene' time interval might put the focus back on understanding Earth's transformation as a long-term social-environmental process. Either way, the causes of Earth's transition to the Anthropocene are human and social. Even as the AWG, led by stratigraphers but including non-geologists too, focuses on defining the Anthropocene as a geological unit following geological criteria, the broader community of social and environmental scientists have every reason to engage in its definition and interpretation, and may need to develop alternative and broader definitions more suited to a deeper historical focus on social-environmental change.

Technosphere

The Anthropocene has also required geologists to embrace new forms of observation and analysis. More than 170,000 'synthetic mineral-like substances', produced only because of human activities, have now been identified, from silicon computer chips to industrial abrasives to ancient ceramics and glass—compared with about 5,000 'natural' minerals. The total volume of Earth now transformed by humans, including the soils altered by agriculture and ocean sediments disturbed by trawling, has also been estimated. The 30 trillion tonne mass of this 'physical technosphere' is truly staggering: 100,000 times greater than Earth's living human biomass (but still only one 200 millionth of Earth's total mass). Plastic materials alone now far exceed human biomass, growing from 2 million tonnes produced annually in 1950 to 300 million in 2015. Total historical production, now 5 billion tonnes, is enough to wrap Earth's entire surface in a thin layer of plastic film.

Geologists have also begun examining the formation of 'technofossils', from cities, roads, and oil rigs to the amazing diversity of manufactured plastic goods, from electronic casings to plastic bottles to microfibres. While the long-term fate of steel girders, electrical wiring, plastics, and many other anthropogenic materials remains uncertain, their potential to fossilize within lake and ocean sediments, landfills, and other stratified deposits, and the sheer amount produced, should more than ensure their survival in the geologic strata. Technofossils also now orbit Earth, rest on its moon and multiple planets, and have even reached interstellar space.

The diversity of 'technospecies' of cultural artefacts might also enable high-resolution timelines of social change to be observed in future strata, in parallel to the assessment of 'material culture' by archaeologists. Already, the diversity of 'technospecies' may exceed the diversity of Earth's 10 million or so living species. Technospecies of electronic gadgets, household goods, and industrial parts is almost certainly in the many millions. Geologists might well begin using 'technostratigraphic markers' to produce geologic time in the future.

Anthroposphere

The Anthropocene has also brought new challenges to Earth system science, including the need to model human systems and the anthroposphere as fundamental components of the Earth system, on a par with its biosphere, atmosphere, and climate systems. Schellnhuber's vision of the Second Copernican Revolution included the anthroposphere within an Earth system 'equation', both as a physical reality and as a metaphysical 'self-conscious control force'—a global human intelligence functioning as a conscious, intentional, 'teleogical', system that would guide Earth towards better outcomes. He was not the first to propose this. Vernadsky himself considered human cognition the 'third stage' in Earth system development, appearing after the geosphere and the biosphere as a globally conscious 'noösphere', based on the concept introduced by the French priest and philosopher Pierre Teilhard de Chardin in the 1920s. Given Earth's current state, one can only wonder what the noösphere is thinking.

Earth system models now incorporate increasingly sophisticated models of human social change and dynamic social-environmental interactions such as shifting economic and agricultural patterns in response to climate change. While it is still early days for these models, interest and investments in this area are growing rapidly.

Earth system science is also working to reconcile its continuous process-based models with the discrete units of stratigraphic time. For example, photosynthesis caused a massive state-shift in the Earth system although it is not recognized by a time interval in the Geologic Time Scale. On the other hand, stratigraphers divide the Holocene from the Pleistocene even though they share the same orbitally forced climate dynamics, making them no different in Earth system terms. The Holocene is just the most recent of dozens of interglacial intervals. On this basis, geologist Ben van der Pluijm and a few others have suggested discarding the Holocene entirely, though the majority of geoscientists regard the Holocene as a useful division. If it were to be removed, the Anthropocene would end the Pleistocene only when human climate forcing overrode orbital forcing as the main driver of Earth's climate dynamics, which had almost certainly occurred by the mid-20th century, and might even have occurred far earlier. From an Earth system perspective, such a shift would not be hard to recognize; Will Steffen estimated anthropogenic warming is now at least 170 times faster than geological background rates based on an 'Anthropocene equation'.

As Schellnhuber predicted, Earth system science is now being asked to provide much more than measurements and predictions. In 2015, the IGBP was transitioned into 'Future Earth', a new international research programme focusing on global sustainability science, in which not only scientists, but also policymakers and business leaders, work together to set the research agenda towards improved environmental governance. Prometheus has indeed been called to Gaia's assistance.

Geoengineering

No form of environmental governance is more Promethean than the 'geoengineering' of Earth's climate. And in the mind of Paul Crutzen, geoengineering and the Anthropocene are deeply entwined. In 2002 he wrote:

There is little doubt in my mind that, as one of the characteristic features of the 'anthropocene', distant future generations of 'homo sapiens' will do all they can to prevent a new ice-age from developing by adding powerful artificial greenhouse gases to the atmosphere. Similarly, any drop in CO_2 levels to too low concentrations, leading to reductions in photosynthesis and agricultural productivity would be combated by artificial releases of CO_2.

Already, there is evidence that anthropogenic greenhouse gas emissions have delayed Earth's next glaciation by 100,000 years. Still, interest in climate geoengineering has never been greater. The more that Earth warms, the more it will cost to deal with the consequences, including disrupted food systems, extended droughts, extreme heat waves, rising seas, severe storms, and other harm to societies. To date, societal efforts are failing to halt rising greenhouse gas emissions. The less that is done now, the more societies will be willing to do for a cooler climate in the future. And geoengineering may turn out to be the surest way to achieve this.

Strategies for climate geoengineering include directly capturing and storing atmospheric CO_2 ('direct air capture'), tree planting, reducing soil tillage, burying charcoal in soils ('biochar'), and fertilizing the oceans, among other enhancements of biological carbon uptake and storage. Alternatively, Earth might be cooled by reflecting the sun's energy back out into space through 'solar geoengineering' ('solar radiation management'), including painting rooftops white and launching giant mirrors into space. Of all the many geoengineering proposals, the most widely discussed, most economically and technologically feasible, and most potentially disruptive, remains Paul Crutzen's 2006 proposal to inject tiny, reflective particles of sulphate aerosols into the stratosphere.

A number of studies, including computer models and estimates of global cooling caused by the 1991 eruption of Mount Pinatubo in the Philippines, have confirmed that sulphate particles sufficient to cool Earth by up to several degrees Celsius could be distributed cost-effectively by a fleet of jet aircraft. Such an intervention might stave off even the highest estimated global warming anticipated by the end of this century (4 to 6 degrees Celsius). Climate scientist David Keith estimated a 1-degree reduction in global temperatures could be sustained by annual expenditures of just 700 million dollars—a tiny sum compared with the costs of eliminating emissions, and within the means of a single nation, or even a single corporation or multi-billionaire.

Despite the temptations of such an inexpensive and actionable 'techno-fix', a sulphate sunshade has genuine potential for catastrophic side effects, from severe droughts to the complete failure of monsoon rains. Allowing atmospheric CO_2 to accumulate might also make the consequences of stopping sulphate injections dramatically worse than any they were intended to prevent. Solar geoengineering using stratospheric sulphates is a shining example of solving one problem by creating an even bigger one. It might still one day be the better option—but without further research, its prospects remain dim.

Icarus

If the Anthropocene were defined only by global climate change, mass extinction, and widespread pollution, that would be enough. Yet these are just a few of the best-known global environmental problems. A single industrial chemical, like the pesticide DDT, had the potential to decimate species around the world. Now, more than 85,000 industrial chemicals are in active use and their production is accelerating (Figure 42). Most have never been tested for harmful effects on humans, let alone other species, or the Earth system as a whole.

Even more concerning is that massively harmful global environmental changes have managed to elude detection for

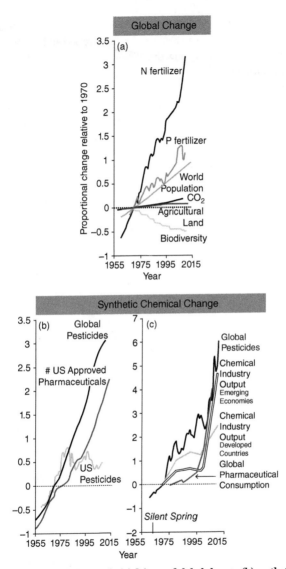

42. **Relative global changes in** (a) **drivers of global change,** (b) **synthetic chemical diversity, and** (c) **synthetic chemical production.**

decades, even when they should have been obvious all along. Two classic examples are ocean acidification (Figure 43) and plastic pollution. You probably know that dissolving carbon dioxide in water makes it more acidic. Yet this simple chemistry was not widely considered a problem for oceans until 2003 when global ecologist Ken Caldeira did the maths that recognized the scale of the threat. Already, the growth of some corals is slowing. If CO_2 emissions continue unabated, Earth's coral reefs and many species of shellfish will disappear by the end of this century, based on comparisons with prior intervals of Earth history with more acidic oceans. Worse still, warmer oceans might do the job first. Yet even in *Planet Under Pressure*, the most comprehensive summary of

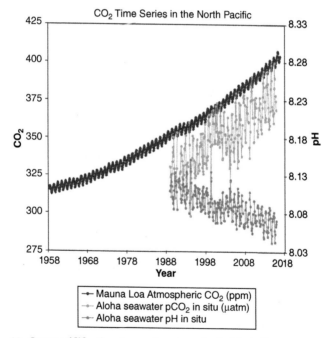

43. Ocean acidification. Increasing atmospheric CO_2 is dissolving in seawater, acidifying (lowering the pH) of the oceans. Measurements off the coast of Hawaii.

Earth system science ever, published in 2004, there was no mention of ocean acidification.

Plastic pollution might seem inevitable considering the incredible tonnage produced. Yet only recently have scientists realized the scale of the problem. Microscopic plastic particles from microfibre clothing, microbeads in cosmetics and cleansers, and the degradation of larger plastic items are accumulating alarmingly in aquatic organisms from microscopic plankton to fish. If trends continue, ocean plastics will outweigh fish by 2050. And these are not the only 'unknown unknowns' of global environmental change. For example, synthetic hormones and other pharmaceuticals are accumulating in freshwater organisms and across entire ecosystems, with relatively unknown consequences.

Given the overwhelming scale, rate, and diversity of harmful global environmental changes produced by human societies, it is hard not to view the Anthropocene as an unmitigated disaster. It might well be viewed as an interval in which humanity, or at least, its wealthiest industrial societies, are driving themselves and the rest of the planet senselessly to ruin. The prospect of a 'bad' Anthropocene defined by toxic environments, declining human health and well-being, war, failed agriculture, submerged cities, catastrophic climate change, mass extinction, and societal collapse, might be unavoidable. Prometheus might be entirely the wrong metaphor. Icarus' foolish hubris to fly in the face of overwhelming odds might prove more accurate. And yet, despite it all, some humans now do in fact fly. It's actually safer than walking down the street.

Good Anthropocenes

The Anthropocene is defined by humans changing Earth so profoundly that they will leave a permanent record in its rocks. Yet there are some who still speak of a 'good Anthropocene'. I myself have been accused of coining the term, but Andrew

Revkin is perhaps the more likely culprit. Either way, I remember first encountering the term at the 2011 Anthropocene meeting in London.

From a scientific point of view, the Anthropocene is neither good nor bad; it is just an observable reality. Yet it should also be clear that the Anthropocene is not yet over. Like other epochs of geologic time it might last for millions of years, with or without us. Better and worse Anthropocenes are real possibilities, depending on what human societies do now and in the future. Moreover, better and worse lower case 'anthropocenes' already exist, depending on how you are experiencing and interpreting this 'age of humans'. You might be living on a low-lying island for example, or you might be the last member of a species on the road to extinction.

To imagine a good Anthropocene is inherently a Promethean act. Yet there are so many ways to be Promethean. Schellnhuber and Crutzen imagined a technocratic Prometheus guided by a sapient 'sphere of human thought', the noösphere, that would use its unprecedented global powers to reverse the environmental harm it had done and to build a better planetary future that might last for millions of years. For otherExt Prometheans, living smaller on this planet might be the game changer—humanity might learn to thrive without transforming Earth, ending the Anthropocene early. There are many other possible anthropocenes between the realms of technocracy and ecotopia and far beyond, from business as usual to a planetary society managed by artificially intelligent robots. But the question remains: are human societies even capable of changing to avoid imminent environmental disaster?

To imagine a good anthropocene it is first necessary to see the better future we have already created. Paul Ehrlich was no fool to predict mass starvation in the 1970s; populations were growing exponentially with no end in sight. Yet rates of human population growth have been declining for decades and more food is now

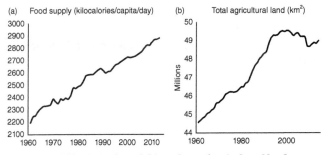

(a) Food supply (kilocalories/capita/day)　(b) Total agricultural land (km²)

44. (a) World food supply, and (b) total use of agricultural land.

produced per person even without significant increases in the global area of land used for agriculture (Figure 44). On average people live longer, healthier, less violent lives, are better educated, and have access to opportunities unimagined by their ancestors. Not only is it possible for human populations to level off, use less land, and live better lives, this is already happening.

Hopes for a technocratic Prometheus are more than just pipe dreams. The Montreal Protocol really did save Earth's ozone layer. There is a long list of societal actions that have staved off environmental disasters, from the banning of DDT and other pollutants, to laws protecting endangered wildlife that have helped bring species back from the brink of extinction. The rise of parks and protected areas, accelerating investments in carbon-neutral energy systems and technologies from solar energy to electric cars, and growth in consumer-driven environmental protections from 'certified sustainable seafood' labelling to 'LEED certified' energy and resource efficient buildings, all raise the prospect of a better planetary future. Future Earth has even invested in a 'Seeds of a Good Anthropocene' project and an *Anthropocene* magazine aimed at identifying and promoting social and environmental innovations that might raise the odds of producing a good Anthropocene. The prospects for anthropocenes much better than the one we are now creating are very real.

Let there be light

In 2014, the Anthropocene entered the *Oxford English Dictionary*, defined as:

> Relating to or denoting the current geological age, viewed as the period during which human activity has been the dominant influence on climate and the environment.

The Anthropocene has entered our lexicon and the scholarly world. Multiple scientific journals now include it in their title.

Does an 'age of humans' mean the end of nature? Have we created a monster? To quote historian of science Bruno Latour:

> Dr. Frankenstein's crime was not that he invented a creature through some combination of hubris and high technology, but rather that he abandoned the creature to itself.

This is not the end of Earth or human history. Conditions will probably support life on Earth for at least another billion years. Our species, like most others, will almost surely be gone by then. Looking back from the deep future, a curious entity might still discover a planet permanently transformed by another.

At this time in which we change the world as we know it, we must also change the way we know the world. The Anthropocene calls on us to think bigger than our individual lives, to imagine the operations of an entire planet and its changes over timescales longer than human societies, from start to finish. It fits in well with broader efforts to reframe education through the lens of 'Big History', a curriculum that connects together historical processes and events from the Big Bang to the present and into the future. It opens us to think forward in deep time, like Stewart Brand and Danny Hillis's Long Now project that is building a clock designed to run for 10,000 years, requiring five-digit years, like 02017.

45. Earth at night. Night-time outdoor lighting detected from space by NASA satellite.

Carl Sagan's *Pale Blue Dot* advised that 'The visions we offer our children shape the future. It *matters* what those visions are. Often they become self-fulfilling prophecies. Dreams are maps.'

Let us look more closely at the 'pale blue dot'. Notice its darker side now glowing vibrantly in the night (Figure 45). It does not glow because anyone ever intended it to. Yet glow it does, by the light of countless human efforts, strung together across the generations and around the world; emergent, social, unintentional. Until this time, the Anthropocene happened while we were busy making other plans. It remains a work in progress.

The Anthropocene tells us that, together, humans are a force of nature. On the road ahead, better and worse anthropocenes exist. The story of the Anthropocene has only just begun. There is still time to shape a future in which both humans and non-human nature thrive together for millennia. There is still a chance for each of us to write a better future into the permanent rock records of Earth history.

Chronology: Potential Anthropocene beginnings, with proposed GSSP markers in bold

Event	Dates	Stratigraphic markers
Stone tools	3.2 million to 2.5 million yr BP	Stone artefacts
Control of fire	1.6 million to 200,000 yr BP	Charcoal
Anatomically modern *Homo sapiens*	~300,000 yr BP	Bones
Behaviourally modern *Homo sapiens*	110,000 to 60,000 yr BP	Complex artefact assemblages, symbolic markings, advanced tools, etc.
Megafauna extinction	50,000 to 10,000 yr BP	Bones, human artefacts, charcoal
Ceramics	30,000 to 15,000 yr BP	Ceramic minerals
Origin of farming	~11,000 yr BP	Pollen (domesticates, weeds), phytoliths, animal bones, charcoal

(continued)

Event	Dates	Stratigraphic markers
Extensive farming	~11,000 to 6,000 yr BP	**~8,000 yr BP CO_2 minima in glacier ice**, pollen (domesticates, weeds), phytoliths, animal bones, charcoal
Rice production, ruminant methane	~6,000 to 3,000 yr BP	**5,020 yr BP CH_4 minima in glacier ice**, animal bones, paddy soils, pollen, phytoliths
Bronze age	~5000 to 3000 yr BP	Metal artefacts, mining, pollution, legacies of deforestation
Biotic homogenization (Homogocene / Homogenocene)	~5000 to 500 yr BP	Pollen, phytoliths, animal bones
Iron age	~3000 to 1,000 yr BP	Iron artefacts, mining, pollution, legacies of deforestation
Anthropogenic soils	~3,000 to 500 yr BP	Soil organic matter, phosphorus accumulations, isotope ratios, pollen
Capitalism (Capitalocene)	~1450	None proposed
Columbian Exchange (Orbis)	1492 to 1610	**1610 CO_2 minima in glacier ice**, pollen, phytoliths, bones, charcoal
Industrial Revolution (Carbocene)	1760 to 1800	Fly ash from coal burning, carbon and nitrogen isotope ratios, diatom composition in lakes, CO_2 in glacier ice.
The Great Acceleration	1945 to 1964	**Radionuclides (1964 ^{14}C & ^{239}Pu peak)**, black carbon, plastics, pollutants, other isotopes

Based in part on Simon L. Lewis and Mark A. Maslin, 'Defining the Anthropocene', *Nature*, 519/7542 (2015), 171–80

References

Chapter 1: Origins

Kolbert, Elizabeth, 'Enter the Anthropocene Age of Man', *National Geographic*, 219/3 (2011), 60–85.

Crutzen, P. J. and Stoermer, E. F., 'The "Anthropocene"', *IGBP Newsletter*, 41 (2000), 17–18.

Revkin, A. C., *Global Warming: Understanding the Forecast* (New York: Abbeville Press, Incorporated, 1992), 180.

Burchfield, Joe D., 'The Age of the Earth and the Invention of Geological Time', *Geological Society, London, Special Publications*, 143/1 (1 January 1998), 137–43.

Arrhenius, Svante, 'On the Influence of Carbonic Acid in the Air upon the Temperature of the Ground', *Philosophical Magazine*, 41 (1896), 237–76.

Report of the Environmental Pollution Panel, President's Science Advisory Committee, 'Restoring the Quality of our Environment' (Washington, DC: The White House, 1965).

Chapter 2: Earth system

Steffen, Will, Crutzen, Paul J., and Mcneill, John R., 'The Anthropocene: Are Humans Now Overwhelming the Great Forces of Nature', *AMBIO: A Journal of the Human Environment*, 36 (2007), 614–21.

NASA Advisory Council. Earth System Sciences Committee, *Earth System Science Overview: A Program for Global Change* (Washington, DC: National Aeronautics and Space Administration, 1986), 48.

Schellnhuber, H. J., '"Earth System" Analysis and the Second Copernican Revolution', *Nature*, 402 (1999), C19–C23.

Moore III, Berrien, et al., 'The Amsterdam Declaration on Global Change', in Will Steffen et al. (eds), *Challenges of a Changing Earth: Proceedings of the Global Change Open Science Conference, Amsterdam, The Netherlands, 10-13 July 2001* (New York: Springer, 2001), 207–8.

Chapter 3: Geologic time

Zalasiewicz, Jan, et al., 'Are we Now Living in the Anthropocene?', *GSA Today*, 18 (1 February 2008), 4–8.

Remane, Jurgen, et al., 'Revised Guidelines for the Establishment of Global Chronostratigraphic Standards by the International Commission on Stratigraphy (ICS)', *Episodes*, 19/3 (1996), 77–81.

International Commission on Stratigraphy, 'Statutes of the International Commission on Stratigraphy, Ratified by IUGS in February, 2002' (International Commission on Stratigraphy, 2002).

Head, Martin J. and Gibbard, Philip L., 'Formal Subdivision of the Quaternary System/Period: Past, Present, and Future', *Quaternary International*, 383 (5 October 2015), 4–35.

Gibbard, Philip L. and Lewin, John, 'Partitioning the Quaternary', *Quaternary Science Reviews*, 151 (1 November 2016), 127–39.

Anthropocene Working Group of the Subcommission on Quaternary Stratigraphy (International Commission on Stratigraphy) '*Newsletter*', No. 1 (2009). Online: <https://quaternary.stratigraphy.org/workinggroups/anthropocene/>.

Zalasiewicz, Jan, Jan, Crutzen, P. J., and Steffen, W., 'The Anthropocene', in F. M. Gradstein et al. (eds), *The Geologic Time Scale 2012 2-Volume Set* (Oxford: Elsevier Science, 2012), 1033–40.

Waters, C. N., et al. (eds), *A Stratigraphical Basis for the Anthropocene* (Geological Society of London Special Publications, Volume 395: Geological Society of London, 2014), 321.

Ruddiman, William F., 'The Anthropogenic Greenhouse Era Began Thousands of Years Ago', *Climatic Change*, 61 (2003), 261–93.

Zalasiewicz, Jan, et al., 'When Did the Anthropocene Begin? A Mid-Twentieth Century Boundary Level is Stratigraphically Optimal', *Quaternary International*, 383 (2015), 196–203.

Chapter 4: The Great Acceleration

Steffen, W., et al., *Global Change and the Earth System: A Planet Under Pressure* (1st edn, Global Change—The IGBP Series; Berlin: Springer-Verlag, 2004), 332.

Steffen, Will, et al., 'The Anthropocene: Conceptual and Historical Perspectives', *Philosophical Transactions of the Royal Society A: Mathematical, Physical and Engineering Sciences*, 369 (13 March 2011), 842–67.

Ellis, Erle C., et al., 'Anthropogenic Transformation of the Biomes, 1700 to 2000', *Global Ecology and Biogeography*, 19 (2010), 589–606.

Smil, Vaclav, *The Earth's Biosphere: Evolution, Dynamics, and Change* (Cambridge, Mass.: MIT Press, 2003), 356.

Vörösmarty, Charles J. and Sahagian, Dork, 'Anthropogenic Disturbance of the Terrestrial Water Cycle', *Bioscience*, 50 (2000), 753–65.

Vitousek, Peter M. and Matson, Pamela A., 'Agriculture, the Global Nitrogen Cycle, and Trace Gas Flux', in Ronald S. Oremland (ed.), *Biogeochemistry of Global Change: Radiatively Active Trace Gases. Selected Papers from the Tenth International Symposium on Environmental Biogeochemistry*, San Francisco, 19–24 August 1991 (Boston: Springer US, 1993), 193–208.

Stocker, Thomas F., et al. (eds), *Climate Change 2013: The Physical Science Basis: A Report of Working Group I of the Intergovernmental Panel on Climate Change* (Cambridge: Cambridge University Press, 2013).

Steffen, Will, et al., 'Stratigraphic and Earth System Approaches to Defining the Anthropocene', *Earth's Future*, 4 (2016), 324–45.

Waters, Colin N., et al., 'The Anthropocene is Functionally and Stratigraphically Distinct from the Holocene', *Science*, 351/6269 (8 January 2016), aad2622.

Chapter 5: Anthropos

Smith, Bruce D. and Zeder, Melinda A., 'The Onset of the Anthropocene', *Anthropocene*, 4 (2013), 8–13.

Marean, Curtis W., 'An Evolutionary Anthropological Perspective on Modern Human Origins', *Annual Review of Anthropology*, 44/1 (2015), 533–56.

Nielsen, Rasmus, et al., 'Tracing the Peopling of the World through Genomics', *Nature*, 541/7637 (01/19/print 2017), 302–10.

Ellis, Erle C., et al., 'Used Planet: A Global History', *Proceedings of the National Academy of Sciences*, 110 (14 May 2013), 7978–85.

Ruddiman, W. F., et al., 'Late Holocene Climate: Natural or Anthropogenic?', *Reviews of Geophysics*, 54/1 (2016), 93–118.

Fuller, Dorian Q., et al., 'The Contribution of Rice Agriculture and Livestock Pastoralism to Prehistoric Methane Levels: An Archaeological Assessment', *The Holocene*, 21 (2011), 743–59.

Boivin, Nicole L., et al., 'Ecological Consequences of Human Niche Construction: Examining Long-Term Anthropogenic Shaping of Global Species Distributions', *Proceedings of the National Academy of Sciences*, 113/23 (6 June 2016), 6388–96.

Lewis, Simon L. and Maslin, Mark A., 'Defining the Anthropocene', *Nature*, 519/7542 (03/12/print 2015), 171–80.

Edgeworth, Matt, et al., 'Diachronous Beginnings of the Anthropocene: The Lower Bounding Surface of Anthropogenic Deposits', *The Anthropocene Review*, 2/1 (8 January 2015), 33–58.

Ruddiman, William F., et al., 'Defining the Epoch we Live in: Is a Formally Designated "Anthropocene" a Good Idea?', *Science*, 348/6230 (2015), 38–9.

Chapter 6: Oikos

Kareiva, Peter, Lalasz, Robert, and Marvier, Michelle, 'Conservation in the Anthropocene', *Breakthrough Journal*, 2 (2011), 26–36.

Comte De Buffon, Georges-Louis Leclerc, *Histoire naturelle générale et particulière: supplement 5: des époques de la nature* (Paris: Imprimerie Royale, 1778). From: Trischler, Helmuth, 'The Anthropocene: A Challenge for the History of Science, Technology, and the Environment', *NTM Zeitschrift für Geschichte der Wissenschaften, Technik und Medizin*, 24/3 (2016), 309–35.

Vitousek, P. M., et al., 'Human Domination of Earth's Ecosystems', *Science*, 277 (1997), 494–9.

Denevan, W. M., 'The Pristine Myth: The Landscape of the Americas in 1492', *Annals of the Association of American Geographers*, 82 (September 1992), 369–85.

Ekdahl, Erik J., et al., 'Prehistorical Record of Cultural Eutrophication from Crawford Lake, Canada', *Geology*, 32/9 (1 September 2004), 745–8.

Zalasiewicz, Jan, et al., 'Making the Case for a Formal Anthropocene Epoch: An Analysis of Ongoing Critiques', *Newsletters on Stratigraphy*, 50/2 (2017), 205–26.

Ceballos, Gerardo, et al., 'Accelerated Modern Human-Induced Species Losses: Entering the Sixth Mass Extinction', *Science Advances*, 1/5 (2015).

Elton, Charles S., *The Ecology of Invasions by Animals and Plants* (London: Butler and Tanner Ltd, 1958), 181.

Vitousek, Peter M., et al., 'Human Appropriation of the Products of Photosynthesis', *BioScience*, 36 (1986), 368–73.

Berkes, F. and Folke, C. (eds), *Linking Social and Ecological Systems: Management Practices and Social Mechanisms for Building Resilience* (Cambridge: Cambridge University Press, 1998), 459.

Sanderson, E. W., et al., 'The Human Footprint and the Last of the Wild', *BioScience*, 52 (2002), 891–904.

Ellis, Erle C. and Ramankutty, Navin, 'Putting People in the Map: Anthropogenic Biomes of the World', *Frontiers in Ecology and the Environment*, 6 (2008), 439–47.

Rockstrom, Johan, et al., 'A Safe Operating Space for Humanity', *Nature*, 461 (2009), 472–5.

Chapter 7: Politikos

Chakrabarty, Dipesh, 'The Climate of History: Four Theses', *Critical Inquiry*, 35 (2009), 197–222.

Crist, Eileen, 'On the Poverty of our Nomenclature', *Environmental Humanities*, 3/1 (1 January 2013), 129–47.

Finney, Stanley C. and Edwards, Lucy E., 'The "Anthropocene" Epoch: Scientific Decision or Political Statement?', *GSA Today*, 26/3–4 (2016), 4–10.

Wilson, E. O., *Half-Earth: Our Planet's Fight for Life* (New York: Liveright, 2016), 256.

Caro, T. I. M., et al., 'Conservation in the Anthropocene', *Conservation Biology*, 26/1 (2011), 185–8.

Hamilton, Clive, 'The Theodicy of the "Good Anthropocene"', *Environmental Humanities,* 7 (2015), 233–8.

Swyngedouw, Erik, 'Apocalypse Now! Fear and Doomsday Pleasures', *Capitalism Nature Socialism*, 24/1 (2013), 9–18.

Scourse, James, 'Enough "Anthropocene" Nonsense: We Already Know the World is in Crisis', *The Conversation* (14 January 2016). <http://theconversation.com/enough-anthropocene-nonsense-we-already-know-the-world-is-in-crisis-43082>.

Boden, T. A., Marland, G., and Andres, R. J., *Global, Regional, and National Fossil-Fuel CO$_2$ Emissions* (2017). Carbon Dioxide Information Analysis Center, Oak Ridge National Laboratory, US Department of Energy (Oak Ridge, Tenn., U.S.A. doi 10.3334/CDIAC/00001_V2017).

Malm, Andreas and Hornborg, Alf, 'The Geology of Mankind? A Critique of the Anthropocene Narrative', *The Anthropocene Review*, 1/1 (1 April 2014), 62–9.

Moore, J. W., *Capitalism in the Web of Life: Ecology and the Accumulation of Capital* (Kindle edn, New York: Verso Books, 2015).

Klein, Naomi, 'Let Them Drown: The Violence of Othering in a Warming World', *London Review of Books*, 38/11 (2016), 11–14.

Biermann, Frank, 'The Anthropocene: A Governance Perspective', *The Anthropocene Review*, 1 (1 April 2014), 57–61.

Heise, Ursula K., 'Terraforming for Urbanists', *Novel*, 49/1 (1 May 2016), 10–25.

Jones, Nicola, 'Solar Geoengineering: Weighing Costs of Blocking the Sun's Rays', *Yale Environment 360* (2014). <http://e360.yale.edu/features/solar_geoengineering_weighing_costs_of_blocking_the_suns_rays>.

Bernhardt, Emily S., Rosi, Emma J., and Gessner, Mark O., 'Synthetic Chemicals as Agents of Global Change', *Frontiers in Ecology and the Environment*, 15/2 (2017), 84–90.

Caldeira, Ken and Wickett, Michael E., 'Oceanography: Anthropogenic Carbon and Ocean pH', *Nature*, 425/6956 (09/25/print 2003), 365–5.

Zalasiewicz, Jan, et al., 'The Geological Cycle of Plastics and their Use as a Stratigraphic Indicator of the Anthropocene', *Anthropocene*, 13 (2016), 4–17.

Bennett, E. M., et al., 'Bright Spots: Seeds of a Good Anthropocene', *Frontiers in Ecology and the Environment*, 14/8 (2016), 441–8.

Latour, Bruno, 'Love your Monsters', *Breakthrough Journal*, 2 (Fall 2011), 21–8.

Sagan, Carl, *Pale Blue Dot: A Vision of the Human Future in Space* (New York: Random House, 1994), 384.

Further reading

Chapter 1: Origins

Carson, Rachel, *Silent Spring* (Houghton Mifflin, 1962), 368.
The classic book that helped spark the environmental movement of the 1960s and beyond.

Leddra, M., *Time Matters: Geology's Legacy to Scientific Thought* (Wiley, 2010). A broad overview of geologic time as a contributor to the sciences.

Mckibben, Bill, *The End of Nature* (Random House, 1989), 226. A hugely influential book about Earth's transformation by anthropogenic climate change.

Marsh, George Perkins, *Man and Nature: or, Physical Geography as Modified by Human Action* (Scribner, 1865). One of the earliest modern books describing dramatic environmental changes caused by human societies.

Schwägerl, Christian and Crutzen, P. J., *The Anthropocene: The Human Era and How It Shapes our Planet* (Synergetic Press, 2014), 248. A solid overview of Anthropocene origins and futures.

Chapter 2: Earth system

Lenton, Tim and Watson, Andrew, *Revolutions that Made the Earth* (Oxford University Press, 2011). An accessible guide to Earth system science as a guide to understanding Earth's functioning today.

Lovelock, J., *Gaia: A New Look at Life on Earth* (Oxford University Press, 1979). The classic book on the Gaia hypothesis.

Vernadsky, Vladimir I., *Biosphere: Complete Annotated Edition*
(Copernicus Books (Springer Verlag), 1998). An annotated
translation of Vernadsky's classic work from 1926.

Chapter 3: Geologic time

Gradstein, Felix M., Ogg, James George, and Schmitz, Mark (eds),
The Geologic Time Scale 2012, 2-volume set (Elsevier, 2012).
The standard text on geologic time. Includes chapters on the
Quaternary, the Prehistoric Human Time Scale, and a new chapter
on the Anthropocene.

Ogg, J. G., Ogg, G., and Gradstein, F. M., *A Concise Geologic Time
Scale: 2016* (Elsevier Science, 2016). An accessible, up-to-date,
book on geologic time by the editors of the standard geology
textbook on geologic time.

Zalasiewicz, Jan, *The Earth After Us: What Legacy Will Humans
Leave in the Rocks?* (Oxford University Press, 2008), 272.

Online: Full listing of approved GSSPs:

International Commission on Stratigraphy: <http://www.stratigraphy.
org/index.php/ics-gssps>

Wikipedia: <https://en.wikipedia.org/wiki/List_of_Global_
Boundary_Stratotype_Sections_and_Points>.

Chapter 4: The Great Acceleration

McNeill, John Robert, *Something New Under the Sun: An
Environmental History of the Twentieth-Century World*
(W. W. Norton & Company, 2001). A classic environmental
history of the 20th century.

McNeill, J. R. and Engelke, P., *The Great Acceleration* (Harvard
University Press, 2016), 288. An environmental history of the
Great Acceleration.

Steffen, W., et al., *Global Change and the Earth System: A Planet
Under Pressure* (1st edn, Global Change: The IGBP Series;
Springer-Verlag, 2004). The classic book defining Earth
system science relating to global environmental change,
including human alteration of the Earth system, featuring an entire
chapter on the Anthropocene. The essential elements of the book
were published earlier, in 2001, as a 32-page summary by the IGBP.

Chapter 5: Anthropos

Henrich, J., *The Secret of our Success: How Culture Is Driving Human Evolution, Domesticating our Species, and Making Us Smarter* (Princeton University Press, 2015). The evolution of exceptional human social abilities.

Mann, Charles C., *1491: New Revelations of the Americas Before Columbus* (Knopf, 2005). A highly readable account of the social and ecological consequences of the Columbian Exchange in the Americas.

Ruddiman, William E., *Plows, Plagues, and Petroleum: How Humans Took Control of Climate* (Princeton University Press, 2005), 224. An excellent book on anthropogenic climate change and climate science in general.

Chapter 6: Oikos

Cohen, Joel E., *How Many People Can the Earth Support?* (W. W. Norton, 1995), 352. The classic book on human populations.

Kareiva, Peter and Marvier, Michelle, *Conservation Science: Balancing the Needs of People and Nature* (Roberts and Company, 2011), 576. A textbook on conserving ecology in the Anthropocene.

Kolbert, E., *The Sixth Extinction: An Unnatural History* (Bloomsbury, 2014), 319. A highly accessible history of species extinction.

Marris, Emma, *Rambunctious Garden: Saving Nature in a Post-Wild World* (Bloomsbury USA, 2011), 224. An exploration of ecology and conservation in the Anthropocene.

Thomas, Chris D., *Inheritors of the Earth: How Nature Is Thriving in an Age of Extinction* (Penguin, 2017), 320. Evolutionary changes in the face of extinction.

Chapter 7: Politikos

Bonneuil, C. and Fressoz, J. B., *The Shock of the Anthropocene: The Earth, History and Us* (Verso Books, 2016), 306. A leftist political history of the Anthropocene.

Davis, Heather and Turpin, Étienne, *Art in the Anthropocene: Encounters Among Aesthetics, Politics, Environments and Epistemologies* (Open Humanities Press, 2015), 416. An edited book exploring the Anthropocene in the Arts.

Haraway, D. J., *Staying with the Trouble: Making Kin in the Chthulucene* (Experimental Futures: Duke University Press, 2016), 312. Donna Haraway's take on the Anthropocene.

Moore, J., et al., *Anthropocene or Capitalocene? Nature, History, and the Crisis of Capitalism* (PM Press, 2016), 240. An edited book with a variety of critical views on the Anthropocene.

Oreskes, Naomi and Conway, Erik M., *Merchants of Doubt: How a Handful of Scientists Obscured the Truth on Issues from Tobacco Smoke to Global Warming* (Bloomsbury Publishing, 2011), 368. A classic on the politics of climate science by an AWG member.

Purdy, J., *After Nature: A Politics for the Anthropocene* (Harvard University Press, 2015), 336. A highly accessible exploration of the political implications of the Anthropocene.

The 'Anthropocene Curriculum' of the Haus Der Kultur De Welt (HKW) is available online here: <http://www.anthropocene-curriculum.org/>.

Chapter 8: Prometheus

Brand, Stewart, *Whole Earth Discipline: An Ecopragmatist Manifesto* (Viking Penguin, 2009), 325. The ultimate promethean expounds on Anthropocene opportunities.

Christian, David and McNeill, W. H., *Maps of Time: An Introduction to Big History* (University of California Press, 2004), 667. Big History connects the contemporary scientific origin story with human history. It is a teaching movement, with free lesson plans online supported by the Big History Project: <https://www.bighistoryproject.com/home>.

Defries, Ruth, *The Big Ratchet: How Humanity Thrives in the Face of Natural Crisis* (Basic Books, 2014), 273. An explanation of why humans transformed Earth's environments and what is to be done about it.

Grinspoon, David, *Earth in Human Hands: Shaping our Planet's Future* (Grand Central Publishing, 2016). A promethean view on the Anthropocene by an astrobiologist.

Morton, Oliver, *The Planet Remade: How Geoengineering Could Change the World* (Princeton University Press, 2015). A powerful book on geoengineering climate.

Pinker, Stephen, *The Better Angels of our Nature: Why Violence Has Declined* (Penguin Publishing Group, 2011), 832. An assessment of long-term declines in violence within human societies.

Shellenberger, Michael and Nordhaus, Ted (eds), *Love your Monsters: Postenvironmentalism and the Anthropocene* (Kindle edn.: Breakthrough Institute, 2011), 102. Promethean assessments of the Anthropocene situation and what to do about it, including the title essay by Bruno Latour.

Online:

Future Earth's *Anthropocene* Magazine: <http://www.anthropocenemagazine.org/>.

Our World in Data <https://ourworldindata.org/:>. Provides useful data and analysis of global social and environmental change.

The Long Now Foundation <http://longnow.org/:>. The Long Now Foundation was established in 1996 to foster long-term thinking and responsibility in the framework of the next 10,000 years.

Anthropocene Journals:

The Anthropocene Review: <http://journals.sagepub.com/home/anr>.

Anthropocene <https://www.journals.elsevier.com/anthropocene/>.

Elementa: Science of the Anthropocene <https://www.elementascience.org/>.

Index

Index

SOCIAL MEDIA
Very Short Introduction

Join our community
www.oup.com/vsi

- Join us online at the official Very Short Introductions **Facebook** page.
- Access the thoughts and musings of our authors with our online **blog**.
- Sign up for our monthly **e-newsletter** to receive information on all new titles publishing that month.
- Browse the full range of Very Short Introductions online.
- Read **extracts** from the Introductions for free.
- If you are a teacher or lecturer you can order inspection copies quickly and simply via our website.

DESERTS
A Very Short Introduction
Nick Middleton

Deserts make up a third of the planet's land surface, but if you picture a desert, what comes to mind? A wasteland? A drought? A place devoid of all life forms? Deserts are remarkable places. Typified by drought and extremes of temperature, they can be harsh and hostile; but many deserts are also spectacularly beautiful, and on occasion teem with life. Nick Middleton explores how each desert is unique: through fantastic life forms, extraordinary scenery, and ingenious human adaptations. He demonstrates a desert's immense natural beauty, its rich biodiversity, and uncovers a long history of successful human occupation. This *Very Short Introduction* tells you everything you ever wanted to know about these extraordinary places and captures their importance in the working of our planet.

www.oup.com/vsi

LANDSCAPES AND GEOMORPHOLOGY

A Very Short Introduction

Andrew Goudie & Heather Viles

Landscapes are all around us, but most of us know very little about how they have developed, what goes on in them, and how they react to changing climates, tectonics and human activities. Examining what landscape is, and how we use a range of ideas and techniques to study it, Andrew Goudie and Heather Viles demonstrate how geomorphologists have built on classic methods pioneered by some great 19th century scientists to examine our Earth. Using examples from around the world, including New Zealand, the Tibetan Plateau, and the deserts of the Middle East, they examine some of the key controls on landscape today such as tectonics and climate, as well as humans and the living world.

www.oup.com/vsi